Cambridge IGCSE™

Physics

STUDY AND REVISION GUIDE

Mike Folland

HODDER
EDUCATION
AN HACHETTE UK COMPANY

Hachette UK's policy is to use papers that are natural, renewable and recyclable products and made from wood grown in well-managed forests and other controlled sources. The logging and manufacturing processes are expected to conform to the environmental regulations of the country of origin.

Orders: please contact Bookpoint Ltd, 130 Milton Park, Abingdon, Oxon OX14 4SB. Telephone: (44) 01235 827720. Fax: (44) 01235 400454. Email: education@bookpoint.co.uk Lines are open from 9 a.m. to 5 p.m., Monday to Saturday, with a 24-hour message answering service. You can also order through our website: www.hoddereducation.com

ISBN: 978 1 4718 5968 7

This text has not been through the Cambridge endorsement process.

The questions, example answers, marks awarded and/or comments that appear in this book were written by the authors. In an examination, the way marks would be awarded to answers like these may be different.

First published in 2005
This edition published in 2016 by
Hodder Education,
An Hachette UK Company
Carmelite House
50 Victoria Embankment
London EC4Y 0DZ

www.hoddereducation.com

Impression number 10 9 8 7 6 5

Year   2020

Cover photo © robertkoczera – Fotolia
Illustrations by Integra Software Services
Typeset by Integra Software Services Pvt. Ltd., India
Printed in India

A catalogue record for this title is available from the British Library.

Contents

REVISED

Introduction

Welcome to the Cambridge IGCSE™ Physics Study and Revision Guide. This book has been written to help you revise everything you need to know for your Physics exam. Following the Physics syllabus, it covers all the key content and provides sample questions and answers, as well as practice questions, to help you learn how to answer questions and to check your understanding.

● How to use this book

> **Main points to revise**
> The key points covered in the chapter.

● Key terms

Definitions of the key terms you need to know.

● Sample questions

Exam-style questions for you to think about.

> **Examiner's tips**
> Advice to help you give the perfect answer.

Student's answers

Typical student answers to see how the question might have been answered.

Examiner's comments

Feedback from an examiner showing what was good, and what could be improved.

> **Now try this**
> Practice questions for you to answer so that you can see what you have learned.

● Answers

Answers to the Now try this questions are found on pages 117–120.

● Extended

Content for the extended syllabus is shaded green.

Tick boxes are provided in the contents page and next to headings so that you can tick off sections as you revise them.

● Exam tips

This revision guide covers the content you need for both the multiple-choice and the written papers. For all questions in both papers, make sure you read the questions carefully and try to answer everything, as you won't lose marks for incorrectly answered questions.

For the multiple-choice paper, work through each question as if it were a normal question – try to avoid guessing, as one of the options is likely to be a 'distractor' that is wrong but at first sight looks likely to be correct. If you cannot answer a question, leave it and return to it at the end when you have answered the other questions.

For the written paper, always make sure you show your working, as it will help you answer the question in a logical way. It will also show the examiner your method. Even if you get the answer wrong, you may still gain marks for using the right method. Always quote the units, label diagrams and give numerical answers to 2 significant figures (s.f.). Finally, use your time wisely. Do not spend too much time on questions that are worth only one mark or write only one line for questions worth a few marks. The Examiner's tips on page 20 give further advice on time management and how much to write.

Cambridge IGCSE Physics Study and Revision Guide Second Edition © Mike Folland 2016

General physics

Main points to revise

- how to measure length, volume and time
- the relationships between distance and speed and how to relate them to moving objects in a variety of situations
- that mass is a measure of the amount of matter in an object
- that weight is a force related to mass
- what is meant by density and how it can be determined
- the effects of forces and how to apply this knowledge to practical situations

- how to describe and carry out an experiment to find the centre of mass of an irregular, flat object
- how to explain the relationship between energy, work and power

- to work out momentum, you need to use the following equation: momentum = mass × velocity ($p = mv$)
- to work out impulse, use the following equation: impulse = force × time = change of momentum ($Ft = mv - mu$)
- the principle of conservation of momentum: when two bodies act on one another, total momentum is conserved
- how pressure is related to force and area
- the relationships between distance, speed, **velocity and acceleration** and how to relate them to moving objects in a variety of situations
- how to explain the distinction between scalars and vectors
- how to draw vector diagrams.

● Key terms

Speed	$\dfrac{\text{total distance}}{\text{total time}}$
Density	$\dfrac{\text{mass}}{\text{volume}}$
Weight	A force exerted by gravity
Moment	The turning effect of a force
Resultant force	Gives an acceleration to an object
Centre of mass	The point where all the mass of an object can be considered to be concentrated
Stability	The ability of an object to return to its original position when disturbed
Momentum	mass × velocity
Impulse	force × time = change of momentum
Renewable	Sources that use energy that can be replaced by natural processes within 50 years
Non-renewable	Sources that use energy that cannot be replaced by natural processes within 50 years
Pollution	The introduction of contaminants into the natural environment
Pressure	$\dfrac{\text{force}}{\text{area}}$
Velocity	Speed in a specified direction
Acceleration	$\dfrac{\text{change of velocity}}{\text{time taken for change}}$
Scalar	A quantity with magnitude (size) only
Vector	A quantity with magnitude (size) and direction
Work	force × distance moved in direction of force
Power	$\dfrac{\text{work done}}{\text{time taken}}$

Measuring length, volume and time

● Length

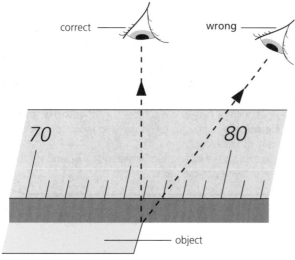

Figure 1.1 The correct way to measure with a ruler

Examiner's tips

• Quote all answers to two significant figures (s.f.), but keep the exact values for intermediate results.

• An answer that is exactly 2 must be written 2.0 to show 2 s.f.

• You must be able to explain how to use a micrometer to measure a very small distance.

• To measure a small thickness, you should measure a number of thicknesses and then divide the total thickness by the number of thicknesses.

Length is the distance from one end of an object to the other. You measure length by looking perpendicularly to the ruler to avoid parallax (see Figure 1.1).

● Volume

Volume is the amount of space occupied. Figure 1.2 shows how to measure volume using a measuring cylinder. You measure the volume of a liquid by looking at the level of the bottom of the meniscus (see Figure 1.2). (For mercury, you should look at the level of the top of the meniscus.)

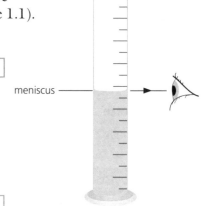

Figure 1.2 The correct way to measure the volume of a liquid

● Common error

✖ Measuring the volume of a liquid from the top of the meniscus.

● Time

You need to be able to use analogue and digital stopwatches or clocks to measure time intervals. A short repeated time interval is measured by timing a number of cycles and then dividing the total time by the number of cycles.

Examiner's tip

Time at least ten cycles and repeat the measurement.

● Common error

✖ Doing the division the wrong way around, that is, calculating the repeated time interval as $\dfrac{\text{number of cycles}}{\text{total time}}$

It is illegal to photocopy this page

Cambridge IGCSE Physics Study and Revision Guide Second Edition © Mike Folland 2016

Now try this

The answers are given on p. 117.

1 A student uses a stopwatch to time the swing of a pendulum. He forgets to zero the timer, which reads 0.5 s when he starts. He starts the stopwatch at the end of the first swing of the pendulum and stops the watch at the end of the tenth swing. The final reading on the timer is 5.9 s. Work out:
  a the number of swings he has timed
  b the time taken for these swings
  c the time for each swing.

Cambridge IGCSE Physics Study and Revision Guide Second Edition © Mike Folland 2016

Speed, velocity and acceleration

Speed

Average speed is calculated using the equation $\dfrac{\text{total distance}}{\text{total time}}$.

Sample question

A runner completes an 800 m race in 2 min 30 s after completing the first lap of 400 m in 1 min 10 s. Find her speed for the last 400 m.   [3 marks]

Student's answer

$$\text{Speed} = \frac{400}{150} = 2.67\,m/s \qquad \text{[2 marks]}$$

Examiner's comments

The student used the correct equation and the correct distance, but used the time for the whole race instead of the time for the last 400 m. The answer is quoted to 3 s.f.

Correct answer

Time = 2 min 30 s – 1 min 10 s = 1 min 20 s = 80 s   [1 mark]

$$\text{Speed} = \frac{400}{80} = 5.0\,m/s \qquad \text{[2 marks]}$$

Examiner's tips
- The equation usually applies only to *constant* acceleration, as it gives the *average* acceleration.
- You must be able to recognise motion for which acceleration is *not* constant.

Velocity is speed in a specified direction. You need to know and be able to use the following equation: acceleration = $\dfrac{\text{change of velocity}}{\text{time}}$.
Deceleration is negative acceleration.

Examiner's tips
- You must be able to recognise linear motion, for which acceleration is constant.

Distance–time graphs show how an object's distance changes with time. The steeper the gradient of a distance–time graph, the greater the speed.

Speed = the gradient of a distance–time graph

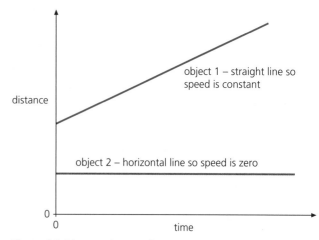

Figure 1.3 Distance–time graph

Cambridge IGCSE Physics Study and Revision Guide Second Edition © Mike Folland 2016

Speed–time graphs show the speed of an object over time. The area under the speed–time graph is the distance covered.

Acceleration occurs when speed changes. The steeper the gradient of a speed–time graph, the greater the acceleration.

A body in free fall near the Earth has constant acceleration, which is often called *g*.

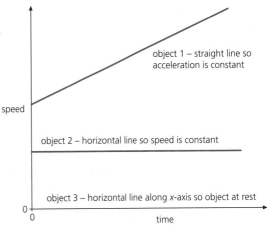

Figure 1.4 Speed–time graph

Acceleration = the gradient of a speed–time graph

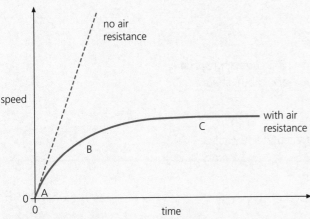

Figure 1.5 A body in free fall in the atmosphere

In the atmosphere there is air resistance. At point A in Figure 1.5, the speed is slow so there is negligible air resistance and the body has free fall acceleration. At point B, the speed is higher and there is some air resistance, so acceleration is less than free fall. At point C, the body has high speed and high air resistance, which is equal to its weight. Therefore, there is no acceleration – this constant speed is called **terminal velocity**.

● Common error

✖ A body in free fall is weightless.
✔ A body in free fall may feel weightless, but weight is the force of the Earth's gravitational field, which still acts on the body.

● Sample question

A car is moving in traffic and its motion is shown in Figure 1.6.

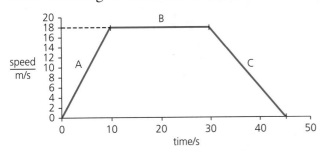

Figure 1.6

1 Choose from the following terms to describe the motion in parts A, B and C: acceleration, deceleration, steady speed. [3 marks]

2 Work out the total distance covered. [5 marks]

3 Work out the acceleration in part C. [2 marks]

Student's answer

1 Part A: acceleration; part B: deceleration; part C: steady speed [1 mark]
2 Distance = speed × time = 18 × 45 = 810 m [0 marks]

3 Acceleration = $\dfrac{\text{change of velocity}}{\text{time}} = \dfrac{18}{15} = 1.2 \text{ m/s}^2$ [1 mark]

Examiner's comments

1 The answers to parts B and C are the wrong way around.

2 The equation used is distance = average speed × time, but this is not appropriate, as the average speed is unknown. The student should have worked out the area under the graph, which equals the distance covered.

3 The calculation is correct but the student should have specified a negative acceleration. [1 mark given]

Correct answers

1 Part A: acceleration; part B: steady speed; part C: deceleration [3 marks]

2 Distance = area under graph [1 mark]

Part A area = $\frac{1}{2}$ × 18 × 10 = 90 m [1 mark]

Part B area = 18 × 20 = 360 m [1 mark]

Part C area = $\frac{1}{2}$ × 18 × 15 = 135 m [1 mark]

Distance = total area = 90 + 360 + 135 = 585 m = 590 m to 2 s.f.

3 Acceleration = $\dfrac{\text{change of velocity}}{\text{time}} = \dfrac{-18}{15} = -1.2 \text{ m/s}^2$ [2 marks]

Now try this
The answers are given on p. 117.

2 A bus accelerates at a constant rate from standstill to 15 m/s in 12 s. It continues at a constant speed of 15 m/s for 8 s.
  a Show this information on a speed–time graph.
  b Use the graph to find the total distance covered.
  c Work out the average speed.

Cambridge IGCSE Physics Study and Revision Guide Second Edition © Mike Folland 2016

Mass and weight

Weight is the force of gravity acting on an object: $W = mg$.
   Mass is the amount of matter in an object.

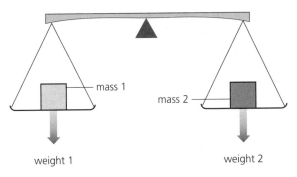

Figure 1.7 Balanced weights

A balance actually compares two weights. As mass determines weight, the balance also compares masses.
   In Figure 1.7, mass 1 = mass 2 because weight 1 = weight 2.

● Experiments to measure density

In order to measure density, you must determine the mass and volume of the material. You need to use the following equation: density $= \dfrac{\text{mass}}{\text{volume}}$.

   For a regularly shaped solid, measure the dimensions and work out the volume, then find the mass on a balance.

   For an irregularly shaped solid, submerge the object in liquid in a large measuring cylinder. The volume of the solid is the increase in the reading (see Figure 1.8). Alternatively, use a displacement can. In this case, the volume of the solid is the volume of liquid displaced.

   For a liquid, measure the volume in a measuring cylinder. To find the mass of the liquid, first find the mass of an empty beaker, pour the liquid into the beaker and then find the total mass of the beaker and the liquid. Work out the mass of the liquid by subtraction of the mass of the beaker from the mass of the total.

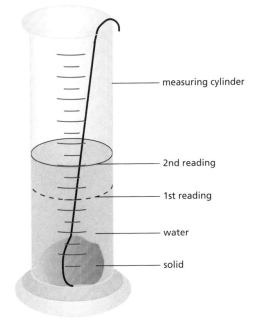

— measuring cylinder

— 2nd reading

— 1st reading

— water

— solid

Figure 1.8 Measuring the volume of an irregular solid

● Sample question

The mass of an empty measuring cylinder is 185 g. When the measuring cylinder contains 400 cm³ of a liquid, the total mass is 465 g. Find the density of the liquid.   [4 marks]

Cambridge IGCSE Physics Study and Revision Guide Second Edition © Mike Folland 2016

Student's answer

$$\text{Density} = \frac{465}{400} = 1.16 \text{ g/cm}^3 = 1.2 \text{ g/cm}^3 \text{ to 2 s.f.}$$ [2 marks]

Examiner's comments

The student put the appropriate quantities into the correct equation and gave the correct units, but used the total mass instead of working out and using the mass of the liquid itself.

Correct answer

Mass of liquid = 465 − 185 = 280 g

$$\text{Density} = \frac{280}{400} = 0.70 \text{ g/cm}^3$$ [4 marks]

> ### Now try this
>
> The answers are given on p. 117.
>
> 3 A measuring cylinder containing 20 cm³ of liquid is placed on a top pan balance, which reads 150 g. More liquid is poured into the cylinder up to the 140 cm³ mark and the top pan balance now reads 246 g. A solid is gently lowered into the cylinder; the liquid rises to the 200 cm³ mark and the top pan balance reads to 411 g. Work out:
> a   the density of the liquid
> b   the density of the solid.

Forces and change of size and shape

Forces

Forces can change the size and shape of a body. You must be able to describe an experiment to measure the extension of a spring, a piece of rubber or another object with increasing load.

Figure 1.9 is an extension–load graph showing the results of such an experiment.

You need to be able to identify and use the term 'limit of proportionality' in extension–load graphs (see Figure 1.9).

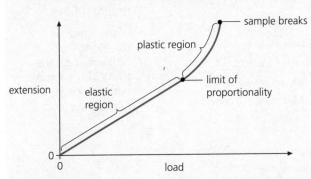

Figure 1.9 Extension–load graph of a typical metal sample loaded to breakage

Hooke's law states that, for the region up to the limit of proportionality, extension is proportional to load. So the graph is a straight line.

You need to be able to recall, understand and use the equation for Hooke's law: $F = kx$.

Forces and change of motion

A **resultant force** gives an acceleration to an object. If the object is stationary, it will gain speed. If the object is moving, it will gain or lose speed depending on the direction of the force.

Friction is a force opposing one surface that is moving or trying to move over another. Air resistance is a form of friction. Friction results in heating.

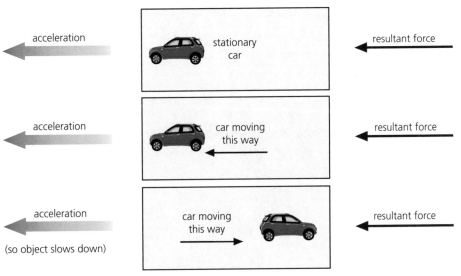

Figure 1.10 A resultant force changes the motion of an object

Cambridge IGCSE Physics Study and Revision Guide Second Edition © Mike Folland 2016

● Common error

✖ If no forward force acts on a moving body, it will slow down.

✔ If a friction force acts on a moving body and there is no forward force, there is a resultant force backwards on the body and it will slow down.

✔ If no resultant force acts on a moving body, it will continue moving with the same speed.

You need to know and be able to use the equation $F = ma$.

F is the resultant force.

Acceleration a is in the direction of the resultant force.

When the force is perpendicular to motion, the object follows a circular path. Some examples of this are shown below.

Object	Force	Circular motion
Planet in orbit	Gravitational force towards the Sun	Planet moves around the Sun
Car turning a corner	Friction force	Car drives around the corner
Ball on a length of string	String tension	Ball whirls around in a circle

Examiner's tips

• You must be able to find the resultant of two or more forces acting in the same line.

• You must state the **direction** of the resultant force.

• In the exam, $g = 10\,\text{m/s}^2$ will be quoted on the front of the paper, not with each question.

● Sample question

An empty lift weighs 2000 N. Four people enter the lift and their total weight is 3000 N. After the button is pressed to move the lift, the tension in the cable pulling up from the top of the lift is 4000 N.

1 Work out the resultant force on the lift. [2 marks]

2 State how the lift moves. [2 marks]

3 Work out the resultant acceleration (take the weight of 1 kg to be 10 N). [4 marks]

Student's answer

1 Resultant force = 3000 + 2000 − 4000 = 1000 N [1 mark]

2 The lift will move down. [1 mark]

3 Mass of lift and people $= \dfrac{5000}{9.81} = 509.7\,\text{kg}$

Acceleration $= \dfrac{f}{m} = \dfrac{1000}{509.7} = 1.962\,\text{m/s}^2$ downwards [3 marks]

Examiner's comments

1 The student correctly worked out the size of the force but did not state the direction downwards.

2 The words 'move down' are too vague.

Cambridge IGCSE Physics Study and Revision Guide Second Edition © Mike Folland 2016

3 The student's answer is correct in itself, but the correctly remembered exact value for *g* was used, not the approximate value quoted, thus making life harder!

The answer is quoted to 4 s.f.

Correct answers

1 Resultant force = 3000 + 2000 − 4000 = 1000 N downwards   [2 marks]

2 The lift will accelerate downwards.   [2 marks]

3 Mass of lift and people = $\dfrac{5000}{10}$ = 500 kg

Acceleration = $\dfrac{F}{m} = \dfrac{1000}{500}$ = 2.0 m/s² downwards   [4 marks]

Now try this

The answers are given on p. 117.

4 A rocket of weight 1000 N is propelled upwards by a thrust of 1800 N. The air resistance is 500 N.
  a Work out the resultant force on the rocket.
  b Describe with an appropriate calculation how this resultant force changes the motion of the rocket.

Cambridge IGCSE Physics Study and Revision Guide Second Edition © Mike Folland 2016

Turning effect and equilibrium

The **moment** of a force is its turning effect.

Moment = force × perpendicular distance from pivot.

An object is in **equilibrium** if there is no resultant turning effect and no resultant force.

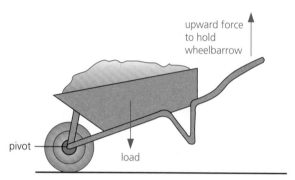

Figure 1.11 Wheelbarrow in equilibrium

Examiner's tips

- You must be able to give some everyday examples of moments or turning effects, e.g. spanner, lever, human arm.

- You must be able to apply moments to the balancing of a beam.

- You must be able to apply moments to new and more complex situations.

● Sample question

A student carries out an experiment to balance a regular 4 m long plank at its mid-point. A mass of 4 kg is placed 80 cm to the left of the pivot and a mass of 3.2 kg is placed 100 cm to the right of the pivot. Explain, *by working out the moments*, whether the plank is balanced.   [4 marks]

Figure 1.12

Student's answer

$4 × 80 = 3.2 × 100$, so the plank balances.   [2 marks]

Examiner's comments

The student's calculation and conclusion are entirely correct, but the instruction in italic to work out the moments was ignored.

Correct answer

Anticlockwise moment = 40 × 0.8 = 32 N m   [1 mark]
Clockwise moment = 32 × 1 = 32 N m   [1 mark]
Anticlockwise moment = clockwise moment, so the plank balances.   [2 marks]

Now try this

The answer is given on p. 117.

5 A see-saw has a total length of 4 m and is pivoted in the middle. A child of weight 400 N sits 1.4 m from the pivot. A child of weight 300 N sits 1.8 m from the pivot on the other side. A parent holds the end of the see-saw on the same side as the lighter child. Work out the magnitude and direction of the force the parent must exert to hold the see-saw level.

Cambridge IGCSE Physics Study and Revision Guide Second Edition © Mike Folland 2016

Centre of mass

A body behaves as if its whole mass were concentrated at one point, called its centre of mass, even though the Earth attracts every part of it. The body's weight can be considered to act at this point.

● Sample question

Explain why the model parrot will only stay on its perch if the bulldog clip is in place. [2 marks]

Student's answer

The bulldog clip lowers the centre of gravity. [1 mark]

Examiner's comments

The centre of mass is lowered but the student did not mention its position relative to the perch. In addition, the student used the old term 'centre of gravity', which is not the correct scientific term.

Correct answer

The bulldog clip moves the centre of mass to directly below the perch, so the parrot is stable. [2 marks]

Examiner's tips

* The lower the centre of mass, the more stable an object becomes.

* Older books may use the term 'centre of gravity', but you should use the correct term, 'centre of mass'.

* You need to be able to describe an experiment to find the centre of mass of a flat object.

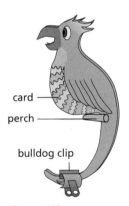

card

perch

bulldog clip

Figure 1.13

Cambridge IGCSE Physics Study and Revision Guide Second Edition © Mike Folland 2016

Scalars and vectors

Scalars:

- only have size (magnitude)
- are added by normal addition.

 Examples of scalars include mass, speed and energy.

Vectors:

- have direction and size (magnitude)
- are added by taking into account their direction.

 Examples of vectors include force, velocity, acceleration and momentum.

Examiner's tips

- You must be able to use a graphical technique to represent two vectors to find their resultant.

- Learn how to use the parallelogram law *or* the triangle law.

- Always quote the scale used in vector diagrams.

● Common errors

✘ Speed is a vector.
✓ Speed is a scalar because it has no direction. Velocity has size *and* direction so it is a vector.
✘ Stating the directions of vectors in ways that apply only to the student's own diagram, e.g. vertically up the paper or assuming that the top of the paper is north.
✓ State the direction of a vector from a definite feature of the situation. (See the sample question below.)

● Sample question

An aircraft flies at 900 km/h heading due south. There is a crosswind of 150 km/h from the west. Graphically, find the aircraft's resultant velocity.

[4 marks]

Student's answer

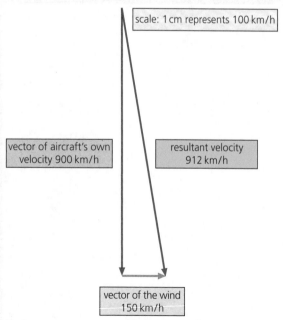

scale: 1 cm represents 100 km/h

vector of aircraft's own velocity 900 km/h

resultant velocity 912 km/h

vector of the wind 150 km/h

Figure 1.14 (Diagram is not drawn to scale)

[3 marks]

Cambridge IGCSE Physics Study and Revision Guide Second Edition © Mike Folland 2016

Examiner's comments

On the whole, the question is extremely well answered and the graphical work is accurate; stating the scale shows excellent work. The student assumed the top of the page was north. However, the student has omitted the direction part of the resultant velocity, stating only the magnitude.

Correct answer

Figure 1.14 should have an arrow pointing up the page labelled 'north'. The answers shown in Figure 1.14 are correct except that the resultant velocity label should be:

resultant velocity 912 km/h at 9° east of due south   [4 marks]

● Sample question   ☐

1 Speed and velocity are related quantities. Explain why speed is a scalar quantity and velocity is a vector.   [2 marks]
2 Name two more scalar quantities and two more vectors.   [4 marks]

Student's answer

1 *Speed is much faster than velocity.*   [0 marks]
2 *Vectors: force, magnetic field strength.*   [2 marks]
  Scalars: energy, colour.   [1 mark]

Examiner's comments

1 The student has shown no understanding of the difference between a scalar and a vector.

2 Two good answers are given as examples of vectors; IGCSE students are not expected to know that magnetic field strength is a vector but it is a correct response.

  Colour is not measurable so is not a scalar quantity.

Correct answers

1 Speed has magnitude only, but velocity has magnitude and direction.   [2 marks]

2 Correct answers could include:

  Vectors: force, acceleration.

  Scalars: energy, mass.   [4 marks]

Cambridge IGCSE Physics Study and Revision Guide Second Edition © Mike Folland 2016

Momentum

Momentum = mass × velocity.
In interaction between bodies, the total momentum is conserved.
Impulse = force × time.
In an interaction, impulse exerted on a body = change of momentum.

● Sample question

A truck of mass 1800 kg moving with a velocity of 4 m/s to the right collides with a truck of mass 1200 kg moving with a velocity of 1 m/s to the left. The truck of mass 1800 kg has a velocity of 1.5 m/s to the right after the collision. Find the final velocity of the 800 kg truck.   [4 marks]

Examiner's tip

Momentum is a vector quantity.

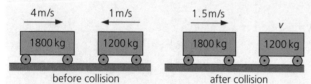

Figure 1.15

Student's answer

Momentum before
collision = 1800 × 4 + 1200 × 1 = 7200 + 1200 = 8400 kg m/s   [0 marks]

Momentum after collision = 1800 × 1.5 + 1200 v = 2700 + 1200 v kg m/s   [1 mark]

Momentum is conserved: 2700 + 1200 v = 8400   [1 mark]

1200 v = 5700

$v = \dfrac{5700}{1200} = 4.75 \, m/s = 4.8 \, m/s$ to 2 s.f.   [1 mark]

Examiner's comments

The student made a good attempt at the question. The working was well set out. However, the student did not realise that direction is significant, as momentum is a vector quantity.

Correct answer

Consider that the positive direction is to the right and assume that v is also to the right.
Momentum before
collision = 1800 × 4 − 1200 × 1 = 7200 − 1200 = 6000 kg m/s   [1 mark]
Momentum after collision = 1800 × 1.5 + 1200 v = 2700 + 1200 v kg m/s   [1 mark]
Momentum is conserved: 2700 + 1200 v = 6000   [1 mark]
1200 v = 3300

$v = \dfrac{3300}{1200} = 2.75 \, m/s = 2.8 \, m/s$ to 2 s.f.   [1 mark]

Now try this

The answers are given on p. 117.

6 A railway truck of mass 6000 kg moving at 6 m/s collides with a truck of mass 10 000 kg moving at 2 m/s in the same direction. The two trucks couple and move on together at 3.5 m/s.
   a Carry out a calculation to confirm that momentum is conserved.
   b Determine if kinetic energy is conserved in the collision.
   c Comment on your answer to part b.

Cambridge IGCSE Physics Study and Revision Guide Second Edition © Mike Folland 2016

Energy

You must understand and be able to identify changes in the forms of energy listed below.

Form of energy	Examples
Kinetic energy or k.e. (energy due to motion)	A car moving; a stone falling; a person running
Gravitational potential energy or g.p.e. (energy due to position)	Water in a mountain lake that can flow downhill to generate electricity; a raised weight that gains gravitational potential energy because it is in a higher position; energy stored in waves and tides that can also be used to generate electricity
Chemical energy released in chemical reactions	Burning fuel to release thermal energy; eating food to provide energy to muscles; providing energy for electrical working in circuits from chemical reactions in a battery; burning fuel in a boiler to provide steam that can drive a turbine to generate electricity
Elastic (strain) energy due to the stretching or bending of materials	Stretching a rubber band; compressing or extending a spring
Nuclear energy released during fission	In nuclear reactors, or fusion as in the Sun
Internal energy (heat or thermal energy)	The increase in temperature when an object is heated
Sound energy	Longitudinal pressure waves that travel through a compressible material
Geothermal energy from within the Earth	Can be used to heat and generate electricity for homes and factories
Light energy and other forms of electromagnetic radiation that can travel through a vacuum	Thermal energy and light from the Sun

Energy can be transferred by forces (mechanical working), electrical currents (electrical working), by heating and by waves.

When energy is transferred, some energy is also transferred to the surroundings as thermal (heat) energy (see Figure 1.16).

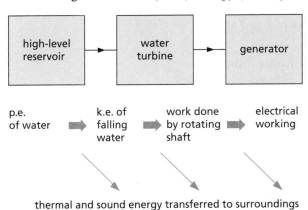

Figure 1.16 Energy transfer also involves some thermal transfer of energy away from the process

Examiner's tips

- The term 'potential energy' includes gravitational and strain energy.

- Energy is *always* conserved. It can be transferred to other forms but cannot be created or destroyed.

- You must be able to give examples of the transfer of energy during events and processes.

- Know and be able to use the following equations:
  k.e. = $\frac{1}{2}mv^2$ and
  g.p.e. = mgh.

It is illegal to photocopy this page

Energy sources

*See 'Examiner's tips' below.

Source	Form of energy	Renewable (R) or non-renewable (NR)	Possible scale*	Expensive*	Air pollution	Other pollution*
Fossil fuels	Chemical	NR	Large	No	Yes	Negligible
Water	Dams g.p.e.	R	Medium	No	No	Loss of habitat
	Tidal g.p.e.	R	Small	No	No	Maritime obstruction
	Waves k.e. and g.p.e.	R	Small	Yes	No	Maritime obstruction
Geothermal	Thermal	R	Medium	No	No	No
Nuclear	Nuclear fission	NR	Large	Yes	No	Radiation, radioactive waste
Radiation from the Sun	Cells, light	R	Small	Yes	No	No
	Panels, thermal	R	Small	Yes	No	
Wind	k.e.	R	Small	Yes	No	Visual, maritime obstruction with offshore schemes
Biomass	Thermal	R	Small	No	Yes	No

Examiner's tips

- There is great variation between different schemes and locations with the properties marked* in the table above. Judgements about these properties are often a matter of opinion.

- You should be able to give the advantages and disadvantages of each method.

Renewable energy sources use energy that can be replaced by natural processes within 50 years.

Energy from the wind, waves and, in some climates, radiation from the Sun can be unreliable. The reliability of energy from hydroelectric dams and radiation from the Sun can be seasonal.

The Sun is the source of all energy resources except geothermal, nuclear and tidal. Energy is released by nuclear fusion in the Sun.

● Sample question

A man winds up the spring of the clockwork radio shown in Figure 1.17 using the muscles in his hand and arm. The internal spring then unwinds to provide energy to power the radio.

Figure 1.17 Winding up a clockwork radio

1 State the form of energy stored in his muscles.   [1 mark]

2 State the form of energy stored in the spring.   [1 mark]

Cambridge IGCSE Physics Study and Revision Guide Second Edition © Mike Folland 2016

3 Name the component that transfers the energy from the spring into useful energy for the radio. [1 mark]

4 Name the process which transfers energy by the electrical currents in the circuits of the radio. [1 mark]

5 Name the form of useful output energy given out by the radio. [1 mark]

6 Most of the energy from the spring will eventually be turned into a form of energy that is not useful. Name this form of energy. [1 mark]

Student's answer

1 Chemical energy is stored in his muscles. [1 mark]
2 Potential energy is stored in the spring. [0 marks]
3 A generator converts the energy in the spring. [1 mark]
4 Electrical working. [1 mark]
5 The useful output energy is radio waves. [0 marks]
6 The waste energy is friction. [0 marks]

Examiner's comments

1 Correct answer.

2 The student should have specified the type of potential energy; a spring stores strain potential energy, often simply called strain energy.

3 Correct answer.

4 Correct answer.

5 The radio is a receiver not a transmitter, so the output is sound energy.

6 Friction is not a form of energy; it is a force which transfers other forms of energy into thermal energy.

Correct answers

1 Chemical energy is stored in his muscles. [1 mark]

2 Strain (potential) energy is stored in the spring. [1 mark]

3 A generator transfers the energy in the spring. [1 mark]

4 The circuits require electrical energy. [1 mark]

5 The useful output energy is sound energy. [1 mark]

6 The waste energy is thermal energy. [1 mark]

● Common error

✖ Energy transferred to wasted thermal energy has been transferred.
✓ Energy can only be transferred, not destroyed. Total energy is always conserved.

Now try this

The answers are given on p. 117. Take $g = 10 \, \text{m/s}^2$.

7 A bungee jumper of mass 60 kg jumps from a bridge tied to an elastic rope that becomes taut after he falls 10 m. Consider the jumper when he has fallen another 10 m and is travelling at 15 m/s.
   a State a form of energy that has been lost.
   b State two forms of energy that have been gained.
   c Work out how much energy is stored in the rope. Ignore air resistance.

Cambridge IGCSE Physics Study and Revision Guide Second Edition © Mike Folland 2016

Examiner's tips

- Some students feel the need to write far too much in answer to descriptive questions, often writing a lot that does not answer the question.

- The answer space on the exam paper is quite sufficient, unless you have exceptionally large handwriting.

- Look at the number of marks available for each part of the question. On average, about one mark is available for each minute of the exam. Descriptive answers may take a little longer, but if you are spending much more time on it than this you are not using your valuable exam time appropriately.

- In this sample question, the student's answer is clearly too short for the number of marks. The correct answer is about the maximum that should be written.

● Sample question

For each of the following two statements, give one strength and one weakness and write a conclusion.

1 A supporter of nuclear power states that it should be more widely used as there is no pollution. [3 marks]

2 A supporter of coal-fired power stations states that nuclear power plants cannot be controlled and might explode like atomic weapons. [3 marks]

Student's answer

1 *Nuclear power stations are constantly leaking radiation, which heats up the atmosphere.* [0 marks]

2 *Nuclear power stations might be damaged in an earthquake or attacked by terrorists. This means there is a constant risk of a serious release of radioactive material.* [2 marks]

Examiner's comments

1 The student has made no real attempt to address the issues or draw a conclusion.

2 This answer is very incomplete but the student has made two relevant points and made a basic conclusion.

Correct answer could be

1 Nuclear power stations produce no carbon dioxide or other air pollution, which is a powerful argument in their favour.

Nuclear power stations produce waste, some highly radioactive, which is very hard to dispose of, as it has a half-life of many centuries. This is a powerful argument against nuclear power.

Because of concerns about the production of greenhouse gases and global warming, the lack of air pollution is a substantial benefit. The issue of radioactive waste is a serious problem. The overall judgement is a matter of opinion with different countries reaching different decisions. [3 marks]

Cambridge IGCSE Physics Study and Revision Guide Second Edition © Mike Folland 2016

2 There have been examples of nuclear power plants overheating causing meltdown of the radioactive core, which has caused considerable release of radioactive substances into the environment. This happened in the Chernobyl incident as a result of an unauthorised test and in Japan following tsunami damage.

However, the statement exaggerates the dangers, as the nuclear material is arranged in a manner that could not cause an explosion like that of a nuclear weapon. In addition, as experience is gained, the likelihood of future incidents decreases.

Again, the two arguments have to be weighed against each other and the overall judgement is a matter of opinion with different countries reaching different decisions. [3 marks]

Now try this

The answers are given on p. 117. Take $g = 10\,\text{m/s}^2$.

8 Supporters and opponents are discussing a proposed new wind farm of 20 large wind turbines. The supporters say that the wind farm will use energy from a renewable source, not pollute and provide cheap, reliable energy. The opponents admit it will use energy from a renewable source, but say that it will be expensive, the energy source will not be reliable and it will pollute.
   a Comment on the arguments of the supporters and the opponents.
   b What is meant by the statement that the wind farm will use energy from a renewable source?
   c Write down one other source of renewable energy and one source of non-renewable energy.

9 a Complete the following sentences about force, work and energy using the terms below.

force acting down the slope due to weight	thermal energy
friction force	time
gravitational potential energy	weight
kinetic energy	work done against friction

A box slides down a steep hill on a road. The box moves at steady speed because the .... equals the .... The loss of .... because of losing height equals .... When the bottom of the box gets hot it gains ....
   b Explain why the electrical power input to an electric motor is greater than the rate that it does work on the output shaft.

10 Complete the table below about devices that transfer energy. The first line has been completed as an example.

Input energy/working	Output energy/working	Device
Chemical	Electrical working	Battery
Electrical working	Light	
Kinetic		Wind turbine
	Electrical working	Solar cell
		Loudspeaker

Cambridge IGCSE Physics Study and Revision Guide Second Edition © Mike Folland 2016

Work, power and efficiency

** Core students should see the 'Examiner's tips' on this page. **

● Work

Work is done when a force moves though a distance. The greater the force, the more work is done. The greater the distance moved, the more work is done. When work is done energy is transferred.

** Work done = force × distance moved in direction of force **

Know and be able to use the following equation: $W = Fd = \Delta E$, where W is work done and ΔE is energy transferred.

Examiner's tip

You must be able to show understanding of work, power and efficiency.

● Power

Power is the rate of doing work. The greater the work done in a given time, the greater the power. The shorter the time in which a given amount of work is done, the greater the power.

$$** \text{Power} = \frac{\text{work done}}{\text{time taken}} ** \text{ OR } ** \text{Power} = \frac{\Delta E}{t} **$$

Examiner's tip

You may find the easiest way to do this is to use the equations marked **, although the syllabus does not actually require you to know these equations.

● Efficiency

The efficiency of a device is the percentage of the energy supplied to it that is usefully transferred.

$$** \text{Efficiency} = \frac{\text{useful energy output}}{\text{energy input}} \times 100\% ** \text{ OR } ** \text{Efficiency} = \frac{\text{useful power output}}{\text{power input}} \times 100\% **$$

● Sample question

The two cranes shown in Figure 1.18 are lifting loads at a port. Crane A raises a load of 1000 N to a height of 12 m in 10 s. Crane B raises the same load of 1000 N to the same height of 12 m but takes 12 s.

Figure 1.18

Cambridge IGCSE Physics Study and Revision Guide Second Edition © Mike Folland 2016

1 Compare, with reasons, the work done by the two cranes.   [2 marks]

2 Compare, with reasons, the power of the two cranes.   [2 marks]

3 Calculate the energy transferred and the power of each crane.   [4 marks]

Student's answer

1 Both cranes do the same amount of work because the force and distance moved are the same.   [1 mark]

2 Crane B has more power because the amount of work done is the same but the time is greater.   [0 marks]

3 Energy transferred by each crane = 1000 × 12 = 12 000
Power of A = 12 000 × 10 = 120 000
Power of B = 12 000 × 12 = 144 000 = 140 000 to 2 s.f.   [1 mark]

Examiner's comments

1 Correct answer.

2 The student has confused the relationship; the shorter the time taken, the greater the power.

3 The calculation of energy transferred is correct, except that the unit (J) has been omitted. Both power calculations are incorrect because the wrong equation has been used; the unit of power (W) has also been omitted.

Correct answers

1 Both cranes do the same amount of work because the force and distance moved are the same.   [2 marks]

2 Crane A has more power because the amount of work done is the same but less time is taken.   [2 marks]

3 Energy transferred by each crane = 1000 × 12 = 12 000 J

$$\text{Power of A} = \frac{12\,000}{10} = 1200 \text{ W}$$

$$\text{Power of B} = \frac{12\,000}{12} = 1000 \text{ W}$$ [4 marks]

● Sample question

The racing car shown in Figure 1.19a is travelling along a straight part of a race track at a constant speed of 95 m/s. The power delivered to the driving wheels is 500 kW.

Figure 1.19a

Cambridge IGCSE Physics Study and Revision Guide Second Edition © Mike Folland 2016

1 Calculate the total resistive force caused by friction and air resistance. [3 marks]

2 Calculate the total work done against this force during the final 600 m of the straight when the car is travelling at constant velocity. [2 marks]

3 On the next lap, a fault develops and the rear wing jams out of position as shown in Figure 1.19b.

The air resistance increases considerably. State and explain any change in the power that would be required to maintain the same speed as in the previous lap. [4 marks]

Figure 1.19b

Student's answer

1 Power = $\frac{\text{work done}}{\text{time taken}}$ = force × $\frac{\text{distance}}{\text{time}}$

$\frac{\text{Distance}}{\text{time}}$ = speed

Force = $\frac{\text{power}}{\text{speed}}$ = $\frac{500\,000}{95}$ = 52 632 N = 53 000 N to 2 s.f.   [2 marks]

2 Work = $\frac{\text{force}}{\text{distance}}$ = $\frac{52\,632}{1000}$ = 53 N   [0 marks]

3 The same car is going at the same speed so there is no change of power.   [0 marks]

Examiner's comments

1 The student has made a calculation error in the final line so loses 1 mark.

2 The student has used the wrong equation for work.

3 The student has not shown any understanding that the increase in resistive force influences the power required.

Correct answers

1 Power = $\frac{\text{work done}}{\text{time taken}}$ = force × $\frac{\text{distance}}{\text{time}}$

$\frac{\text{Distance}}{\text{time}}$ = speed

Force = $\frac{\text{power}}{\text{speed}}$ = $\frac{500\,000}{95}$ = 5263 N = 5300 N to 2 s.f.   [3 marks]

2 Work = force × distance = 5263 × 600 = 3.1579 × 10⁶ N = 3.2 × 10⁶ N to 2 s.f.   [2 marks]

3 The correct answer would include:
- the total resistive force would increase
- the work done for every metre moved along the straight would increase
- at the same speed, the work done every second would increase
- the power required would increase.   [4 marks]

> ## Now try this
>
> The answer is given on p. 117. Take $g = 10\,\text{m/s}^2$.
>
> 11 In a small-scale hydroelectric power scheme, 24 kg of water falls every second through a vertical height of 60 m from the reservoir to the turbine. The electrical output is 11 kW. Find the efficiency of the scheme.

Cambridge IGCSE Physics Study and Revision Guide Second Edition © Mike Folland 2016

Pressure

To work out pressure you need to use the following equation: $p = \dfrac{F}{A}$

The unit of pressure is the pascal (Pa). A force of 1 N on an area of 1 m² exerts a pressure of 1 Pa.

● Common error

✖ Carelessly using the word 'pressure' instead of 'force' in the answer to descriptive questions.

A barometer is used to measure atmospheric pressure in millimetres of mercury, which is given by the height of the mercury column (Figure 1.20). Space Y is a near-vacuum containing a small amount of mercury vapour.

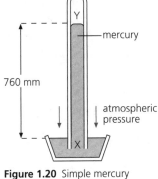

Figure 1.20 Simple mercury barometer

● Liquid pressure

Pressure beneath a liquid surface depends on the depth and the density of the liquid. The greater the depth in a given liquid, the greater the pressure. At a given depth, the greater the density of the liquid, the greater the pressure.

You should know and be able to use the following equation: $p = h\rho g$, where h is the depth below the surface of the liquid, ρ is the density of the liquid and g is the acceleration due to gravity.

The manometer shown in Figure 1.21 is used to compare the pressure of a gas with atmospheric pressure. The difference in pressure is measured by the height difference (h) between the two columns.

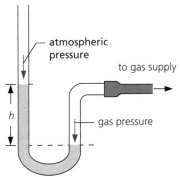

Figure 1.21 Manometer or U-tube

● Sample question

Some students are playing a ball game in the sea and the ball is pushed 60 cm below the surface of the water (see Figure 1.22). (Density of seawater = 1.025 × density of freshwater.)

1 Compare, with reasons, the pressure on a point on the ball 60 cm below the surface of the sea with the pressure just below the surface. [2 marks]

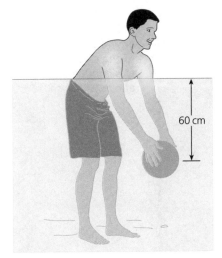

Figure 1.22

Cambridge IGCSE Physics Study and Revision Guide Second Edition © Mike Folland 2016

2 Compare, with reasons, the pressure on the ball 60 cm below the surface of the sea with the pressure 60 cm below the surface of a freshwater lake. [2 marks]

3 Calculate the pressure on a point on the ball 60 cm below the surface of the sea (density of freshwater = 1000 kg/m³; take g = 10 m/s²). [2 marks]

Student's answer

1 The pressure increases. [1 mark]
2 The pressure on the ball below the surface of the sea is greater because seawater has a greater density. [1 mark]
3 Pressure = $h\rho g$ = 0.6 × 1000 × 10 = 6000 Pa [1 mark]

Examiner's comments

1 The statement is correct but no reason is given.

2 The statement is correct and the reason is also correct, but not quite complete. The student should have mentioned that the comparison was at the same depth below each surface.

3 The pressure has been calculated in the correct way but at a depth of 60 cm below the surface of freshwater instead of seawater.

Correct answers

1 The pressure increases because the ball is at a greater depth in the same liquid. [2 marks]

2 The pressure on the ball below the surface of the sea is greater because seawater has a greater density and both balls are at the same depth. [2 marks]

3 Density of sea water = 1.025 × 1000 kg/m³ = 1025 kg/m³

Pressure = $h\rho g$ = 0.6 × 1025 × 10 = 6150 Pa = 6200 Pa to 2 s.f. [2 marks]

Now try this

The answers are given on p. 117.

12 An engineer uses the mercury manometer shown in Figure 1.23 to test the pressure of a gas supply. Write down at which of the points (A, B or C):
  a the pressure is greatest
  b the pressure is smallest.

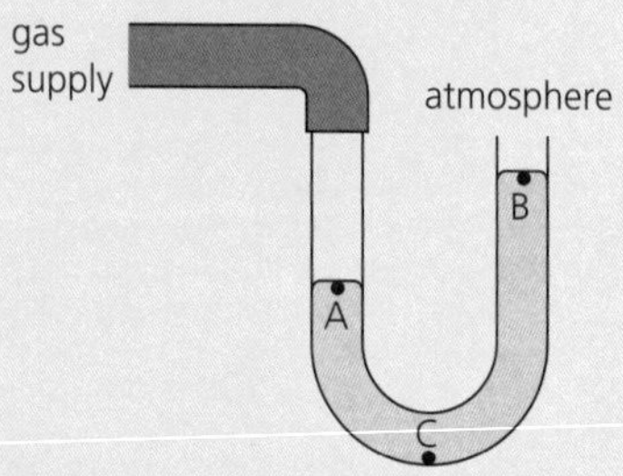

gas supply   atmosphere

B

A

C

Figure 1.23

Examiner's tip

Keep all the figures of an intermediate answer. See the first line of the correct answer to sample question 3 above with 4 s.f.

Cambridge IGCSE Physics Study and Revision Guide Second Edition © Mike Folland 2016

● Sample question

An aeroplane passenger buys a plastic bottle of water during a flight. She drinks most of the water and closes the cap. Figure 1.24 shows the bottle as she puts it in her bag. After landing, she sees that the bottle has changed shape.

1 In the space to the right of Figure 1.24, draw the new appearance of the bottle. **[1 mark]**

2 Explain why the bottle has this new shape. **[2 marks]**

Student's answer

1

Figure 1.24

[0 marks]

2 *As the aeroplane descends, the pressure outside the bottle decreases, but the pressure inside the bottle stays the same, which causes the bottle to blow up.* [1 mark]

Examiner's comments

The student has incorrectly stated that atmospheric pressure decreases as the aeroplane descends. However, the explanation that follows on from this wrong assumption is logical.

Correct answers

1

Figure 1.24

[1 mark]

2 As the aeroplane descends, the pressure outside the bottle increases but the pressure inside the bottle stays low. This causes the bottle to collapse. **[2 marks]**

Cambridge IGCSE Physics Study and Revision Guide Second Edition © Mike Folland 2016

TOPIC 2 Thermal physics

Main points to revise

- how the kinetic theory explains the nature of solids, liquids and gases
- how to explain the behaviour of solids, liquids and gases in a wide range of practical situations from everyday life
- how temperature is measured
- how the flow of thermal energy affects the properties of solids, liquids and gases and their changes of state
- how thermal energy is transferred by conduction, convection and radiation.

● Key terms

Term	Definition
Molecule	A tiny particle consisting of one, two or more atoms
Particle	Any small piece of a substance; it could be one molecule or billions of molecules
Temperature	A measure of the amount of internal or thermal energy within a body
Thermal energy	Energy that flows into or out of a body by conduction, convection or radiation
Thermal capacity	The amount of thermal energy needed to increase the temperature by 1 °C
Specific heat capacity	The thermal energy needed per kilogram to increase the temperature of a material by 1 °C
Specific latent heat	The thermal energy needed to change the state of 1 kg of material
Conduction	When thermal energy is transferred through a material without movement of the material
Convection	When thermal energy is transferred due to movement of liquid or gas
Radiation of thermal energy	When thermal energy is transmitted by electromagnetic radiation

Cambridge IGCSE Physics Study and Revision Guide Second Edition © Mike Folland 2016

Kinetic theory of molecules

All matter is made up of molecules in motion. The higher the temperature, the faster the motion of the molecules. Almost always, matter expands with increases in temperature.

Solids

Key features of solids:

- Molecules are close together.
- Molecules vibrate about fixed points in a regular array or lattice.
- The rigid structure of solids results from these fixed positions.
- As temperature increases, the molecules vibrate further and faster. This pushes the fixed points further apart and the solid expands.

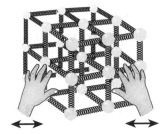

Figure 2.1 A model of the molecular behaviour of a solid

- There is only a very slight expansion of a solid with increases in temperature, e.g. the length of an iron rod increases by about 0.1% when it is heated from 20°C to 100°C.
- The positions of molecules in a solid are fixed because the attractive and repulsive forces between neighbouring molecules are balanced.

Liquids

Key features of liquids:

- Molecules are slightly further apart than in solids.
- Molecules are still close enough to keep a definite volume.
- The main motion of the molecules is vibration. The molecules also move randomly in all directions, not being fixed to each other.
- As temperature increases, the molecules move faster and further apart, so the liquid expands. One exception to this is that, when liquid water is heated from 0°C to 4°C, its structure changes, so it contracts instead of expands.

- The forces between molecules are too weak to keep them in a definite pattern but are sufficient to hold them to the bulk of the liquid.

- There is a small expansion of a liquid with increases in temperature, e.g. the volume of many liquids increases by about 4% when heated from 20°C to 100°C.

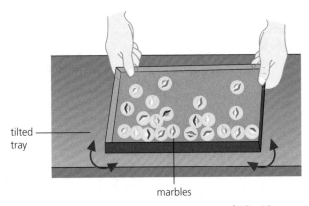

tilted tray

marbles

Figure 2.2 A model of the molecular behaviour of a liquid

Cambridge IGCSE Physics Study and Revision Guide Second Edition © Mike Folland 2016

Gases

Key features of gases:

- Molecules are much further apart than in solids or liquids.
- Molecules move much faster than in solids or liquids.
- There is no definite volume. Molecules move throughout the available space.
- Molecules constantly collide with each other and the container walls.
- Gases have low densities.
- The higher the temperature, the faster the speed of the molecules. In fact, temperature is a measure of the average speed of the molecules.
- The higher the temperature, the larger the volume of a gas at constant pressure.

- There is a considerable expansion of a gas with increases in temperature at constant pressure, e.g. the volume of a gas increases by about 27% when it is heated from 20 °C to 100 °C.

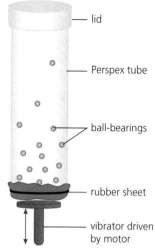

Figure 2.3 A model of the molecular behaviour of a gas

lid
Perspex tube
ball-bearings
rubber sheet
vibrator driven by motor

Brownian motion

In the apparatus in Figure 2.4a, smoke particles reflect the light, which is seen in the microscope as tiny bright dots. They move around haphazardly, as in Figure 2.4b. They also appear and disappear as they move vertically. This movement is caused by the irregular bombardment of the smoke particles by fast-moving, **invisible** air molecules. This is clear evidence for the kinetic theory.

Kinetic theory was first observed by Robert Brown, who observed in a microscope pollen particles suspended in water moving haphazardly due to bombardment by fast-moving, invisible water molecules.

Smoke/pollen particles are visible and relatively massive.
Air/water molecules are invisible and fast moving.

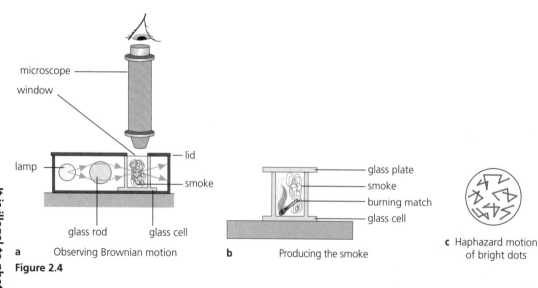

microscope
window
lamp
lid
smoke
glass rod
glass cell

a Observing Brownian motion

glass plate
smoke
burning match
glass cell

b Producing the smoke

c Haphazard motion of bright dots

Figure 2.4

Common error

✘ Incorrect use of the words 'particle' and 'molecule' shows a lack of understanding of their different properties and roles.

Cambridge IGCSE Physics Study and Revision Guide Second Edition © Mike Folland 2016

● Sample question

A student looks in a microscope at a cell containing illuminated smoke particles. Explain:

1 what is seen

2 the movement observed

3 what causes this movement.   [4 marks]

Student's answer

1 *Smoke particles.*   [0 marks]
2 *Moving around.*   [0 marks]

3 *The smoke molecules are bombarded by air.*   [1 mark]

Examiner's comments

1 It is reflected light, not smoke particles, that is seen.

2 'Moving around' is too vague.

3 The word 'molecule' is incorrect and the whole answer is incomplete.

Correct answers

1 Bright specks of light.   [1 mark]

2 Moving around haphazardly in *all* directions.   [1 mark]

3 The bright specks are light reflected off the smoke particles, which are bombarded by air molecules.   [2 marks]

Cambridge IGCSE Physics Study and Revision Guide Second Edition © Mike Folland 2016

Gas pressure

Gas pressure is caused by the total force of collisions between fast-moving molecules and the walls of the container they are in. The higher the temperature, the faster the molecules move. If the volume is kept constant, the pressure increases because:

- there are more frequent collisions with the container walls
- the collisions are harder, so exert more force.

The change of momentum of the molecules when they strike the walls creates a force.

You should be able to explain gas pressure in terms of the change of momentum of the molecules at each collision.

Now try this

The answers are given on p. 117.

1 A gas molecule strikes the wall of a container and bounces back. Explain:
  a in terms of momentum, how this causes a force on the walls of the container
  b how all the molecules of the gas cause a pressure on the walls of the container.

● Common error

✖ Mentioning collisions between molecules when explaining gas pressure. It is true that the molecules collide with each other, but this does not explain gas pressure.

● Gas pressure and volume at constant temperature

At a **constant temperature**, gas molecules move at a constant average speed, so the average force from each collision is the same. If the gas is compressed into a smaller volume, there are more frequent collisions on each unit of area, so the total force per unit area increases and the pressure increases.

Similarly, if the gas expands to a greater volume at a constant temperature, the pressure decreases.

You should know and be able to use the equation for a fixed mass of gas at constant temperature:

pressure × volume = constant ($p_1 V_1 = p_2 V_2$)

Examiner's tip

If you set out your working logically, as shown in the worked example, you are much more likely to get the answer right.

● Worked example

Question

A piston slowly compresses a gas from $540\,cm^3$ to $30\,cm^3$, so that the temperature remains constant. The initial pressure was $100\,kPa$; find the final pressure.

Cambridge IGCSE Physics Study and Revision Guide Second Edition © Mike Folland 2016

Solution

$p_1 = 100 \text{ kPa}$, $p_2 = \text{unknown}$

$V_1 = 540 \text{ cm}^3$, $V_2 = 30 \text{ cm}^3$

$p_1 V_1 = p_2 V_2$

$100 \times 540 = p_2 \times 30$

$p_2 = \dfrac{100 \times 540}{30} = \dfrac{54\,000}{30} = 1800 \text{ kPa}$

● Sample question

A gas cylinder is heated in a fire. State what happens to the pressure of the gas and explain your answer in terms of the molecules. [4 marks]

Student's answer

The pressure increases because the molecules move around more, hitting each other and the walls. [1 mark]

Examiner's comments

The student's answer is vague, mentioning the molecules colliding with each other, which is irrelevant.

Correct answer

The pressure increases because the molecules move faster, [2 marks] hitting the walls more frequently and harder, thus increasing the total force on the walls. [2 marks]

Now try this

The answers are given on pp. 117–118.

2 An experiment is carried out on some gas contained in a cylinder by a piston, which can move.
In stage 1, the gas is heated with the piston fixed in position. State and explain whether the following will increase, decrease or stay the same during stage 1:
a the speed of the molecules
b the number of collisions per second between molecules and the walls
c gas pressure.
In stage 2, the gas stays at a fixed temperature while the piston moves to increase the volume of the gas. State and explain whether the following will increase, decrease or stay the same during stage 2:
a the speed of the molecules
b the number of collisions per second between molecules and the walls
c gas pressure.

It is illegal to photocopy this page

Expansion of solids, liquids and gases

In the fire alarm circuit in Figure 2.5, thermal energy from the fire causes the lower metal in the bimetallic strip to expand more than the upper metal. This causes the strip to curl up, which completes the circuit and the alarm bell rings.

Uses of expansion of a solid: shrink-fitting, curling of a bimetallic strip in a fire alarm.

Disadvantage of expansion of a solid: gaps need to be left between lengths of railway line to allow for expansion in hot weather.

Use of expansion of a liquid: liquid-in-glass thermometers; see the later section 'Measurement of temperature'.

Disadvantage of expansion of a liquid: the water in a car's cooling system expands when the engine gets hot. A separate water tank is needed for the hot water to expand into.

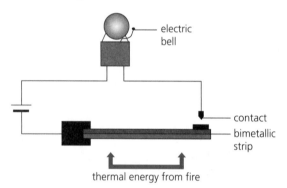

Figure 2.5 A fire alarm

● Sample question

The lid is stuck on a glass jar. How could you use hot water to release it? Explain in terms of the molecules how this works. [4 marks]

Student's answer

Put the glass jar in hot water and the lid will come off [1 mark] because the molecules expand. [0 marks]

Examiner's comments

The student did not specify where exactly the hot water should be used and gave a vague, incorrect explanation of the role of the molecules.

Correct answer

The correct answer should include:
Put the lid in hot water so that it expands and can be released. [2 marks] The molecules in the lid will move faster; their mean positions move further apart, so the lid expands. [2 marks]

Cambridge IGCSE Physics Study and Revision Guide Second Edition © Mike Folland 2016

Measurement of temperature

To measure temperature or changes in temperature, a physical property is needed that varies in a regular way over a wide range of temperatures. A practical thermometer is a simple piece of apparatus that measures how such a property changes.

Examples of properties that can be used: expansion of a solid (coiled bimetallic strip) or liquid (mercury or alcohol); electrical voltage between two junctions of different metals (thermocouple); electrical resistance.

● Common errors

✖ Confusing thermal energy (heat) and temperature.
✓ Thermal energy flows from a hot body to a cold body.
✓ Temperature measures the amount of thermal or internal energy in a body. In everyday terms, it measures how hot a body is.

Sensitivity means the amount the property changes for each degree the temperature changes. For sufficient sensitivity, a coiled bimetallic strip thermometer, for example, must have enough coils to give sufficient movement and a liquid-in-glass thermometer, as another example, must have a narrow tube or large bulb.

Range means the difference between the highest and lowest temperatures a thermometer can measure. The range must suffice for the intended use, e.g. a thermometer for domestic use would have a range from 0 °C to 50 °C.

Linearity means the reading must change by the same amount for every degree of temperature change, e.g. a property that changes little in one half of the temperature range and much more in the other half would not be suitable.

The **fixed points** of a thermometer are essential to give it its scale. The thermometer must read exactly 0 °C at the freezing point of pure water and exactly 100 °C at the boiling point of pure water at normal atmospheric pressure. In between these fixed points, the scale is divided into equal divisions.

● Liquid-in-glass thermometers

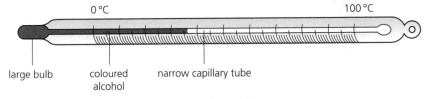

0 °C   100 °C

large bulb   coloured alcohol   narrow capillary tube

Figure 2.6 An alcohol thermometer with a Celsius scale

Mercury and alcohol are suitable liquids because:

● they expand enough to make a sensitive thermometer
● they can be used over wide ranges of temperatures – alcohol from −115 °C to 78 °C (and higher if under pressure) and mercury from −39 °C to 357 °C
● they both expand linearly with increases in temperature
● Because of safety concerns, many countries have banned the sale of mercury thermometers to the general public.

Cambridge IGCSE Physics Study and Revision Guide Second Edition © Mike Folland 2016

In liquid-in-glass thermometers, sensitivity is increased by having a large bulb, a narrow capillary tube or a liquid which expands more. Range is increased by having a longer capillary tube or a less sensitive thermometer. Linearity is achieved by using a suitable liquid and ensuring the bore of the capillary tube is constant.

● Sample question

An alcohol-in-glass thermometer used to measure temperatures in a house has a range from –5 °C to 50 °C. A technician designs a thermometer to be used in a laboratory to measure temperatures between –10 °C and 100 °C.

1 State why a red dye is added to the alcohol in the thermometers.   [1 mark]

2 Compare the ranges of the thermometers.   [1 mark]

3 State and explain two ways in which the house thermometer could be redesigned for use in the laboratory.   [4 marks]

Student's answer

1 *So it shows.*   [1 mark]

2 *It is higher.*   [0 marks]
3 *Bigger capillary so there is more room.*   [0 marks]
  Small bulb so it does not expand so much.   [1 mark]

Examiner's comments

The wording of this student's answers is too imprecise and too short. Students should not feel they need to write a lot, but the answer should be sufficient to give the necessary information.

1 Not a good answer but enough to score the mark.

2 'Higher' could mean either a larger range, which is correct, or the same range at higher values. In addition, the student has not said which thermometer the comment applies to. Answers will score marks only when it is clear that they are correct.

3 'Bigger capillary' might mean longer or greater diameter.
Small bulb is correct. The expansion comment could refer to the volume of expansion or movement along the capillary.

Correct answers

1 Red dye is added to make the transparent alcohol visible.   [1 mark]

2 The laboratory thermometer has a larger range.   [1 mark]

3 The correct answer could include any two of the following:
  • longer stem of the capillary tube, so the liquid can keep expanding above 50 °C
  • larger diameter of the capillary, so the liquid does not expand so much for each degree rise in °C
  • smaller bulb, so not as much liquid moves to the capillary for each degree rise in °C.   [4 marks]

Cambridge IGCSE Physics Study and Revision Guide Second Edition © Mike Folland 2016

Now try this

The answers are given on p. 118.

4 A mercury-in-glass thermometer is placed in pure melting ice and the mercury bead is 12 mm long.
   a State the Celsius temperature that the thermometer should read.
   The thermometer is now placed in steam above boiling water and the bead expands to 82 mm long.
   b State the Celsius temperature that the thermometer should now read.
   c Work out the length of the bead at 50 °C.
   d Work out the temperature reading when the bead length is 61 mm long.

5 A clinical thermometer is used to measure the temperature of the human body. It has a range of 35 °C to 42 °C. For each of the following properties, explain how it compares with a normal domestic thermometer, the reason for any difference and how it is achieved by the design of the clinical thermometer:
   a sensitivity
   b range
   c linearity.

● Thermocouple thermometer

The voltage produced between the two junctions of wires of different metals is proportional to the temperature difference between the junctions. Only one junction is attached to the object whose temperature is being measured. This junction is very small and light, so reacts quickly to rapidly changing temperatures. It is also very robust and resistant to damage from vibrations of machinery. Thermocouples are widely used in industry because they can be used over a wide temperature range, from below −250 °C to 1500 °C. In addition, the voltage output is very convenient for recording data; an industrial test rig may have hundreds of thermocouple thermometers all connected to computers for analysis.

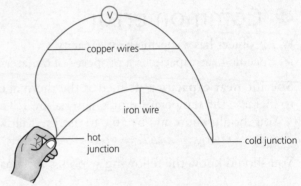

Figure 2.7 A simple thermocouple thermometer

● Sample question

A new petrol engine is being tested and the engineers need to measure the temperature of the exhaust pipe close to the engine. State, with reasons, the type of thermometer that would be used. [4 marks]

Student's answer

A thermocouple, because it is small and goes up to a high temperature. [2 marks]

Examiner's comments

The student's answer is on the right lines but rather vague and incomplete. (It would earn the first and last marks below.)

Correct answer

A thermocouple would be used [1 mark] because the hot junction is small [1 mark] and will not be damaged by the vibrations of the engine. [1 mark] A thermocouple can measure high temperatures. [1 mark]

Cambridge IGCSE Physics Study and Revision Guide Second Edition © Mike Folland 2016

Thermal capacity and specific heat capacity

When thermal energy flows into a body, its molecules move faster, increasing its internal energy and its temperature. The **thermal** (or heat) **capacity** of a body is the amount of thermal energy needed to increase its temperature by 1 °C. The thermal capacity is determined by the mass of the body and its material.

● Common error

✖ A material has a thermal (or heat) capacity.
✔ A body has a thermal (or heat) capacity.

** Core students should see 'Examiner's tips' on this page. **

You should know and be able to use the following equation:
** change of energy = thermal capacity × temperature change **

● Common error

✖ An object has a specific heat capacity.
✔ Specific heat capacity is a property of a material.

Specific heat capacity is defined as the thermal energy needed *per kilogram* to increase the temperature of a material by 1 °C.

You should know and be able to use the following equation:
change of energy = $mc\Delta T$.

You should know the following symbols: m = mass, c = specific heat capacity, ΔT = temperature change.

Examiner's tips

- **Core students** must be able to show understanding of thermal capacity.

- **Core students** may find the easiest way to do this is to use the equation marked **, although the syllabus does not actually require you to know this equation.

● An experiment to measure the specific heat capacity of a metal

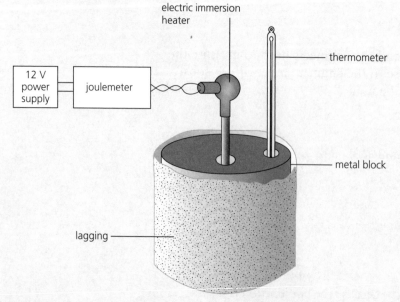

Figure 2.8 The lagging reduces thermal energy transfer to the surroundings

Cambridge IGCSE Physics Study and Revision Guide Second Edition © Mike Folland 2016

Measure the mass of the metal block and the temperature before and after heating, and record the joulemeter reading of the energy supplied (Figure 2.8). Use these results to calculate the specific heat capacity of the block.

Example of working out:

Mass of metal block = 1.6 kg

Temperature before heating = 21 °C

Temperature after heating = 66 °C

Increase of temperature = 45 °C

Joulemeter reading = 46 800 J

$$\text{Specific heat capacity} = \frac{\text{energy supplied by immersion heater}}{\text{mass} \times \text{temperature increase}} = \frac{46\,800}{1.6 \times 45}$$

$$= 650 \text{ J/(kg °C) to 2 s.f.}$$

Now try this

The answers are given on p. 118.

6 An experiment is carried out to find the specific heat capacity of a metal. A 2 kg block of the metal is heated by a 200 W heater for 5 min, and the temperature of the block rises from 20 °C to 51 °C. Work out:

a the energy supplied to the block by the heater

b the specific heat capacity of the metal.

When used in an engine, a component made from this metal receives 35 kJ of thermal energy and its temperature rises from 30 °C to 290 °C.

c Work out the thermal capacity of the component.

d Work out the mass of the component.

Cambridge IGCSE Physics Study and Revision Guide Second Edition © Mike Folland 2016

Changes of state

Evaporation occurs when molecules at the surface of a liquid, which are moving fastest, have enough energy to escape the attractive force of the rest of the liquid and become molecules of a gas. Evaporation takes place at all liquid temperatures.

Energy is needed to break the bonds between molecules, so evaporation causes the molecules in the remaining liquid to cool down.

The rate of evaporation increases with:

- higher temperatures, as more molecules at the surface are moving faster

- increased surface area, as more molecules are at the surface

- a wind or draught, as the gas molecules are blown away so cannot re-enter the liquid.

● Sample question

A student is playing football on a cool, windy day, wearing a T-shirt and shorts. He feels comfortably warm because he is moving around vigorously. His kit then gets wet in a rain shower. Explain why he now feels cold. [2 marks]

Student's answer

The wet T-shirt makes him feel cold. [0 marks]

Examiner's comments

The student's answer is far too vague and does not mention the cooling caused by evaporation.

Correct answer

The water in his wet kit is evaporated by the wind. [1 mark] The thermal energy needed for this evaporation is taken from the water in his T-shirt and shorts, as well as from his body, so he feels cold. [1 mark]

Boiling occurs at a definite temperature called the 'boiling point'. Bubbles of vapour form within the liquid and rise freely to the surface. Energy must be supplied continuously to maintain boiling.

Condensation occurs when gas or vapour molecules return to the liquid state. Thermal energy is given out as the bonds between molecules in the liquid re-form.

Melting, or fusion, takes place at a definite temperature called the 'melting point'. Thermal energy must be provided to break the bonds between molecules for them to leave the well-ordered structure of the solid.

Solidification, or freezing, occurs when molecules of a liquid return to the solid state. This takes place at a definite temperature called the 'freezing point', which has the same value as the melting point. Thermal energy is given out as the bonds between molecules of the solid re-form.

Examiner's tip

You must be able to distinguish between boiling and evaporation. Carefully learn the features of each.

Cambridge IGCSE Physics Study and Revision Guide Second Edition © Mike Folland 2016

Now try this

The answers are given on p. 118.

7 An ice cube with a temperature of 0 °C is placed in a glass of water with a temperature of 20 °C. After a few minutes, some of the ice has melted. State and explain whether the following increase, decrease or stay the same:
  a the temperature of the remaining ice
  b the temperature of the water
  c the mass of the water
  d the total mass of the ice and water.

8 Write down three differences and two similarities between boiling and evaporation.

● Common error

✖ Temperature increases during melting and boiling because thermal energy is being supplied.

✔ Temperature stays constant during melting and boiling. The thermal energy supplied is used to break the bonds between molecules.

● Specific latent heat

Energy must be provided to change the state from a solid to a liquid or a liquid to a gas. This energy is called 'latent heat' and is used to break bonds between molecules, not to increase temperature.

Specific latent heat is defined as the thermal energy needed per kilogram to change the state of a material, at constant temperature.

You need to be able to use the equation: energy = ml.

You should know the symbols: m = mass, l = specific latent heat.

An experiment to measure the specific latent heat of vaporisation of water (steam)

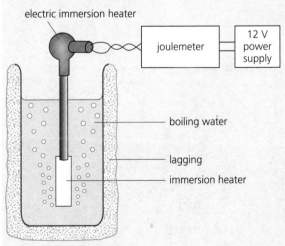

Figure 2.9 The lagging and tall, narrow shape reduce thermal energy losses to the surroundings

Measure the mass of the water and its container before and after the experiment (Figure 2.9). Bring the water to the boil before starting the joulemeter. Record the joulemeter reading of the energy supplied during boiling. Use these results to calculate the specific latent heat of vaporisation.

Example of working out:

Initial mass of water and container = 650 g

Final mass of water and container = 580 g

Mass of water vaporised by immersion heater

= initial mass of water and container – final mass of water and container

= 650 g – 580 g = 70 g = 0.07 kg

Joulemeter reading = 245 000 J

$$\text{Specific latent heat of vaporisation} = \frac{\text{energy supplied}}{\text{mass of water vaporised}}$$

$$= \frac{245\,000}{0.07}$$

$$= 3\,500\,000 \text{ J/kg}$$

● Sample question

Ice is melted in a lagged funnel by an immersion heater supplied with electricity through a joulemeter. The melted water is collected in a beaker. Use the data supplied to work out the specific latent heat of fusion of ice.

Joulemeter readings:

before experiment 148 000 J

after experiment 172 600 J

Mass of empty beaker: 90 g

Mass of beaker with melted water: 150 g [4 marks]

Student's answer

Heat emitted by immersion heater = 172 600 – 148 000 = 24 600 J [1 mark]

$$\text{Specific latent heat of fusion} = \frac{24\,600}{0.150} = 164\,000 \text{ J/kg}$$ [1 mark]

Examiner's comments

The student has made a good attempt at the calculation but has forgotten to subtract the mass of the beaker itself.

Correct answer

The correct answer includes:

Thermal energy emitted by immersion
heater = 172 600 – 148 000 = 24 600 J [1 mark]

Mass of ice melted = 150 – 90 = 60 g = 0.060 kg [1 mark]

$$\text{Specific latent heat of fusion} = \frac{24\,600}{0.060} = 410\,000 \text{ J/kg}$$ [2 marks]

Cambridge IGCSE Physics Study and Revision Guide Second Edition © Mike Folland 2016

Conduction, convection and radiation

Thermal energy is always transferred from a place of high temperature to a place of low temperature.

● Transfer by conduction

In **conduction**, thermal energy is transferred through a material without movement of the material.

Metals are generally good conductors, but most other solids are poor conductors. Liquids are generally much worse thermal energy conductors than metals.

Gases are all very poor conductors of thermal energy. For example, if you put your hands in very cold water, they will feel cold almost at once. If your hands are in air of the same temperature, they will cool down but at a much slower rate because air is a bad conductor.

The atoms in a hot part of a solid vibrate faster and further than those in a cold part.

In metals, thermal energy is transferred by fast-moving free electrons, which pass through the solid, causing atoms in colder parts to vibrate more.

There is a secondary mechanism that is much slower. The vibrating atoms cause their neighbours to vibrate more, thus passing on thermal energy. Non-metals do not have free electrons so can use only this mechanism, which is why non-metals are poor conductors.

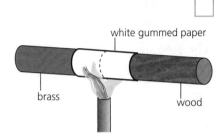

Figure 2.10 The paper over the brass does not burn because brass is a good conductor

Figure 2.11 Water is a poor conductor of thermal energy

● Transfer by convection

In **convection**, thermal energy is transferred owing to movement of the liquid or gas itself.

The liquid or gas expands on heating, so its density falls. The warmer and lighter liquid or gas rises to the cooler region, transferring thermal energy in the process.

Figure 2.12 demonstrates a convection current. The heated air above the candle rises up the left-hand chimney and draws smoke from the lighted paper down the right-hand chimney and into the box.

Convection currents in water can be seen by dropping a potassium permanganate crystal into a beaker of water. The coloured traces indicate the flow of the convection currents.

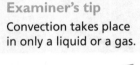

Examiner's tip
Convection takes place in only a liquid or a gas.

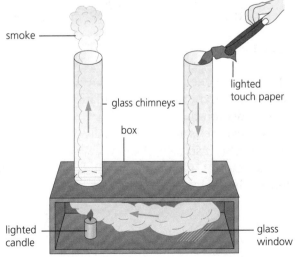

Figure 2.12 Demonstrating convection in air

43

● Transfer by radiation

In **radiation**, thermal energy is transferred by infrared radiation, which is part of the electromagnetic spectrum.

Surfaces that are good absorbers of thermal energy radiation are also good emitters. Black surfaces are better emitters and absorbers than white surfaces. Dull or matt surfaces are better emitters and absorbers than polished surfaces.

There need not be any matter between the hot and cold bodies (no medium is required).

Most solids and liquids absorb infrared radiation, including water, which is transparent to light.

The amount of radiation emitted increases with the surface area and surface temperature.

Examiner's tip

Core students do not have to describe the two experiments below, but if you understand them it will help you to answer questions on the core syllabus.

Good and bad emitters of infrared radiation

The copper sheet has previously been heated strongly with a Bunsen burner. The hand next to the black surface feels much hotter than the hand next to the polished surface. This is because black surfaces emit thermal energy radiation more than polished surfaces.

hot copper sheet with one side polished and the other blackened

back of hands towards sheet

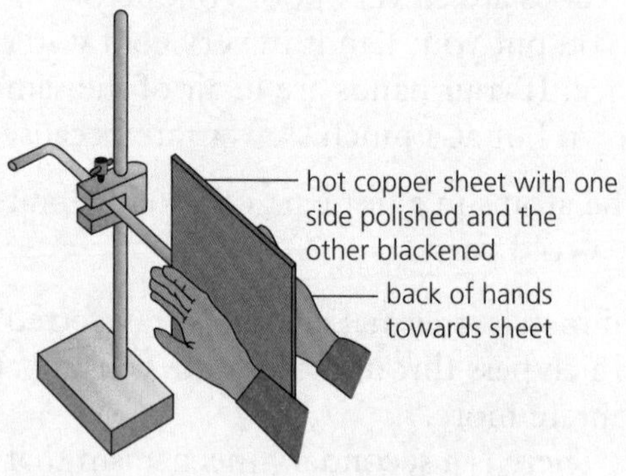

Figure 2.13 Comparing emitters of radiation

Good and bad absorbers of infrared radiation

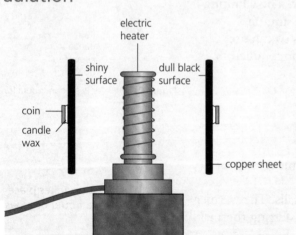

Health and safety

Figure 2.13 Safety note: if you do this experiment, the copper plate needs to be very hot. If touched, it could cause a serious burn.

Figure 2.14 For clarity, the essential protective guard round the heater has not been shown. This safety feature is essential to protect users from a hot object at high electrical voltage.

electric heater

shiny surface

dull black surface

coin

candle wax

copper sheet

Figure 2.14 Comparing absorbers of radiation

The heater is the same distance away from each copper sheet, so each receives the same amount of radiation. The dull black surface absorbs much more radiation than the shiny surface, so after a few minutes the wax on the black sheet melts and the coin falls off. The shiny surface reflects a lot of the radiation, so it stays cool and the wax does not melt.

The amount of radiation depends on the surface temperature and surface area of the body.

● Everyday uses and consequences of the transfer of thermal energy

Examiner's tip

There are countless everyday uses and consequences of the transfer of thermal energy. As a minimum, you should learn and be able to explain the following examples. If possible, read up and understand many more, especially coastal breezes, the vacuum flask and the greenhouse.

Saucepans and other solids through which thermal energy must travel are made of metals such as aluminium or copper, which are **good conductors**.

Blocks of expanded polystyrene are used for house insulation because they contain trapped air, which is a **bad conductor**.

A domestic radiator heats the air next to it which then rises and transfers thermal energy to the rest of the room. Despite its name, a radiator works mainly by **convection**.

Double glazing reduces the transfer of thermal energy by trapping a narrow layer of air between the window panes and **reducing convection**.

The Sun heats the Earth by **radiation** through space.

Refrigerators have cooling pipes at the back. These have fins to give a larger surface area and increase loss of thermal energy by convection and radiation. The fins are also painted black to increase thermal energy loss by radiation because black surfaces are **good emitters**.

Many buildings in hot countries are painted white because white surfaces are **bad absorbers** of radiation from the Sun.

● Common error

✖ Black surfaces increase thermal energy transfer by conduction and convection.
✓ The colour of the surface influences radiation only.

● Sample question

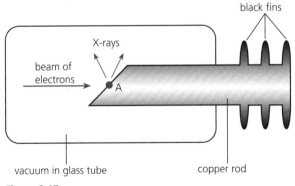

Figure 2.15

Figure 2.15 shows an X-ray tube. Only a small proportion of energy from the electrons that strike point A goes into the X-rays that are emitted from that point. Most of the energy is transferred to thermal energy at point A; this energy is removed by the copper rod. Explain how conduction, convection and radiation play a role in the removal of this thermal energy.   [6 marks]

Student's answer

The thermal energy goes down the rod and is conducted out by the fins [1 mark] through convection and radiation.   [1 mark]

Examiner's comments

The student's answer shows an incorrect understanding of conduction.

Correct answer

Thermal energy is **conducted** along the rod to the fins [2 marks] and is then emitted from the fins to the air by **convection** and **radiation**. [2 marks] The black colour of the fins increases the rate of radiation. [2 marks]

Now try this

The answers are given on p. 118.

9 A boy goes for a walk in winter in a cold country.
   a He opens a metal gate which makes his hands cold. State and explain which type of thermal energy transfer cools his hands.
   b He then washes his hands in a stream. State and explain which type of thermal energy transfer cools his hands in this case.
   c He comes across some workmen who have lit a fire to keep themselves warm and he holds out his hands towards the fire. State and explain which type of thermal energy transfer warms his hands.

Waves

Main points to revise

- the basic nature of wave motion
- how to distinguish between transverse and longitudinal waves
- how to describe how water waves can be used to illustrate reflection, refraction and diffraction
- how to draw diagrams of light rays undergoing reflection and refraction and passing through thin converging lenses
- that refractive index $n = \dfrac{\text{speed of wave in first material}}{\text{speed of wave in second material}}$

- the main features of the electromagnetic spectrum
- how to use the wave equation: $v = f\lambda$
- how wavelength and gap size affect diffraction through a gap
- how to describe that, with increasing wavelength, there is less diffraction at an edge
- how to use the following equation: $\dfrac{\sin i}{\sin r} = n$.

● Key terms

Term	Definition
Transverse wave	Travelling wave in which oscillation is perpendicular to the direction of travel
Longitudinal wave	Travelling wave in which oscillation is parallel to the direction of travel
Speed of a wave	The distance moved by a point on a wave in 1 s
Frequency	The number of complete cycles that occur in 1 s
Wavelength	The distance between corresponding points in successive cycles of a wave
Amplitude	The maximum displacement of a wave from the undisturbed position
Refractive index	Refractive index = speed of wave in medium 1/speed of wave in medium 2
Real image	An image that can be formed on a screen by the intersection of rays
Virtual image	An image seen by observing rays diverging from it
Converging lens	A lens which refracts parallel rays of light such that they converge to meet at a point
Principal focus	Rays of light parallel to the principal axis are refracted by the lens to pass through the principal focus
Focal length	The distance between the principal focus, F, and the optical centre, C
Electromagnetic spectrum	Waves of the same nature with a wide range of wavelengths made up of oscillating electric and magnetic fields
Compression	Regions where particles of material are closer together
Rarefaction	Regions where particles of material are further apart

Cambridge IGCSE Physics Study and Revision Guide Second Edition © Mike Folland 2016

Waves

Waves transfer energy from one point to another without transferring matter. Some waves (e.g. water waves and sound waves) are transmitted by particles of a material vibrating about fixed points. They cannot travel through a vacuum.

Electromagnetic waves (e.g. light waves and X-rays) are a combination of travelling electric and magnetic fields. They *can* travel through a vacuum.

In **transverse waves**, the oscillation of the material or field is at right angles to the direction of travel of the wave. Figure 3.1 shows a transverse wave travelling in a horizontal rope. Each piece of rope oscillates vertically about a fixed point, but the pieces do not oscillate in time with each other.

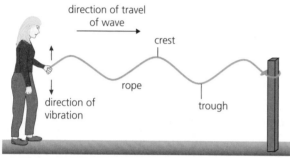

Figure 3.1 Transverse waves in a rope

● Common error

✖ Transverse waves oscillate vertically when the wave travels vertically.
✓ Transverse waves oscillate horizontally when the wave travels vertically because the oscillation is always at right angles to the direction of travel.

In **longitudinal waves**, the oscillation of the material is parallel to the direction of travel of the wave.

Figure 3.2 represents a longitudinal wave travelling in a horizontal spring. Each coil of the spring oscillates horizontally about a fixed point, but the coils do not oscillate in time with each other. C marks the points where the coils are most tightly packed (**compressions**) and R marks the points where the coils are furthest apart (**rarefactions**).

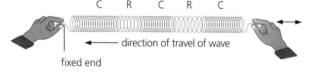

Figure 3.2 Longitudinal waves in a spring

● Common error

✖ Longitudinal waves oscillate horizontally when the wave travels vertically.
✓ Longitudinal waves oscillate vertically when the wave travels vertically, because the oscillation is always parallel to the direction of travel.

The speed (v) of a wave is the distance moved by a point on the wave in 1 s.

The **frequency** (f) of a wave is the number of complete cycles that occur in 1 s and is measured in hertz (Hz).

Cambridge IGCSE Physics Study and Revision Guide Second Edition © Mike Folland 2016

The **wavelength** (λ) of a wave is the distance between two corresponding points (e.g. crests) in successive cycles.

The **amplitude** of a wave is the maximum displacement of the wave from the undisturbed position (marked a in Figure 3.3).

Examiner's tips
- You need to be able to use the term 'wavefront' as a line showing the position of a wave.

- The wavefront shows similar points (e.g. crests) of an extended travelling wave, such as a wave in water.

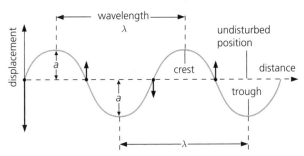

Figure 3.3 Displacement–distance graph for a wave at a particular instant

Speed, frequency and wavelength are related by the equation $v = f\lambda$.

● Sample question

Sketch one and a half cycles of a transverse wave and mark on your sketch the amplitude and wavelength. [4 marks]

Student's answer

The student's answer is shown by dashed lines in Figure 3.4. [1 mark]

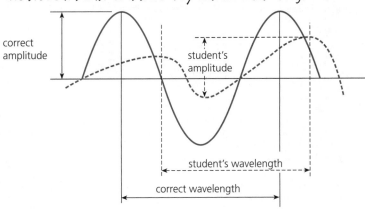

Figure 3.4

Examiner's comments

With this type of question, it is essential to work carefully and accurately or few marks will be gained. Even for a sketch, the student's diagram is too casual. For minimum acceptable accuracy, the student must indicate that each half wavelength is the same length, and the distance from the axis to each crest and trough is the same. The student's amplitude is incorrectly measured from crest top to trough bottom. Given the irregular wave, the student has correctly labelled one wavelength.

Correct answer

The correct answer is shown by the solid red line in Figure 3.4. [4 marks]

● Common error

✖ Amplitude is the height difference between the top of a crest and the bottom of a trough.

✔ Amplitude is the height difference between the top of a crest and the mean position, or between the bottom of a trough and the mean position.

Cambridge IGCSE Physics Study and Revision Guide Second Edition © Mike Folland 2016

● Sample question

A sensor detects that 1560 cycles of a wave pass in 30 s. Work out the
frequency of the wave.   [3 marks]

Student's answer

$Frequency = 52 \text{ cycles in } 1 \text{ s.}$ [2 marks]

Examiner's comments

The student's answer is correct but the unit of frequency is hertz (Hz).
Note that, although it is a fairly easy calculation, there is no working. If the student had made
a slight slip, no credit could have been given for using the correct method.

Correct answer

$$Frequency = \frac{number\ of\ cycles}{time} = \frac{1560}{30} = 52\ Hz$$ [3 marks]

● Sample question

Find the frequency of a radio wave with a wavelength of 1500 m.
(Please note, extended candidates are expected to know the speed of
electromagnetic waves.)   [3 marks]

Student's answer

$v = f\lambda$ [1 mark]

$So\ f = \frac{v}{\lambda} = \frac{3 \times 10^8}{\lambda} = 200\ 000\ kHz$ [1 mark]

Examiner's comments

The student has done everything correctly, but the unit is wrong. It should be Hz not kHz.
Perhaps the student assumed that, as radio frequencies are often expressed in kHz, this
was the correct unit.

Correct answer

$v = f\lambda$ [1 mark]

$So\ f = \frac{v}{\lambda} = \frac{3 \times 10^8}{\lambda} = 200\ 000\ Hz = 200\ kHz$ [2 marks]

Now try this

The answers are given on p. 118.

1 A woman swimming in the sea estimates that when she is in the trough between two crests of
   a wave, the crests are 1.5 m above her.
   a Work out the amplitude of the wave.
   An observer counts that the swimmer moves up and down 12 times in 1 min.
   b Work out the frequency of the wave.

Cambridge IGCSE Physics Study and Revision Guide Second Edition © Mike Folland 2016

Summary of some types of waves:

Type of wave	Longitudinal/transverse	Travel through a material
Wave on a rope	Transverse	Material needed
Wave on a spring	Either	Material needed
Water	Transverse	Material needed
Earthquake wave	Both	Material needed
Sound	Longitudinal	Material needed
Electromagnetic, e.g. light, X-rays, radio waves	Transverse	No material needed, but some electromagnetic waves can travel through certain materials

● Sample question

An earthquake wave is travelling vertically down into the Earth; the oscillations are also vertical. State, with a reason, whether the wave is longitudinal or transverse. [2 marks]

Student's answer

The wave is transverse because it is vibrating up and down. [0 marks]

Examiner's comments

It is the direction of oscillation relative to the direction of travel that matters – the student does not mention this.

Correct answer

The wave is longitudinal because the oscillations are parallel to the direction of travel. [2 marks]

Now try this

The answers are given on p. 118.

2 A child throws a ball into a pond, hears the sound of the splash and observes water waves travelling towards him.
   a As the sound waves travel towards him, in which direction are the air particles oscillating?
   b As the water waves travel towards him, in which direction are the water particles oscillating?

● Water waves

We can observe a wave travelling on the water surface of a ripple tank to illustrate how waves behave.
   Key features of a ripple tank:

- A beam just touching the surface vibrates vertically to produce a wave.
- A light source shines through the water and shows the wave pattern on a screen above or below the ripple tank.

Examiner's tip

You must be able to describe how water waves can be used to show reflection, refraction and diffraction.

Reflection

In Figure 3.5, the wave produced by the vibrating beam is reflected from the flat metal barrier. The reflected wave is at the same angle to the reflecting surface as the incident wave. Speed, wavelength and frequency are unchanged by reflection.

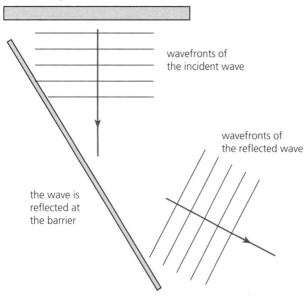

Figure 3.5 Reflection of a wave in a ripple tank

Refraction

Figure 3.6 shows a wave entering the shallow water above the glass where its speed is reduced. The frequency stays the same, so the wavelength is also reduced. The refracted wave changes direction.

Diffraction from a narrow gap

Speed, wavelength and frequency are unchanged by diffraction, as shown in Figure 3.7.

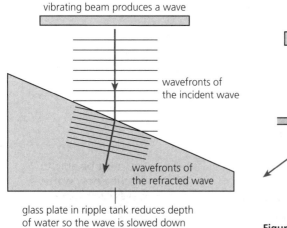

Figure 3.6 Refraction of a wave in a ripple tank

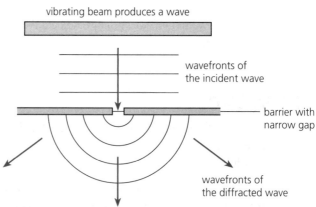

Figure 3.7 Diffraction of waves in a ripple tank by a gap that is narrower than the wavelength

Cambridge IGCSE Physics Study and Revision Guide Second Edition © Mike Folland 2016

● Common errors

✖ Carelessness in drawing diffraction diagrams.
✖ Poor semi-circles not drawn with the middle of the narrow gap as their centre.
✖ Different wavelengths of incident and diffracted waves.
✖ Varying wavelengths of the diffracted wave.
✔ To draw the diffracted waves accurately, measure the wavelength of the incident waves. This is the radius of the first circle. For each subsequent circle, the increase of radius must be measured to be the same as the wavelength.
✔ Careful, accurate measuring and drawing is *essential* to produce good diagrams.

Diffraction from a wide gap

With increasing wavelength, there is less diffraction at an edge, as shown in Figure 3.8.

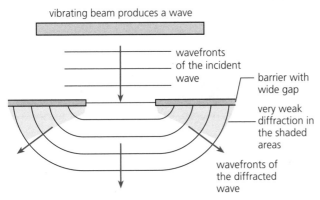

Figure 3.8 Diffraction of waves in a ripple tank by a gap that is greater than the wavelength

Examiner's tip

The centres of the quarter-circles must be at the edges of the wide gap.

Cambridge IGCSE Physics Study and Revision Guide Second Edition © Mike Folland 2016

Light

Light moves as waves of very small wavelength, but it is often convenient to use light rays to work out and explain the behaviour of light.

A light ray is the direction in which light is travelling and is shown as a line in a diagram.

An object is what is originally observed.

An optical image is a likeness of the object, which need not be an exact copy.

A real image is formed where the rays cross and can be shown on a screen.

A virtual image is observed where rays appear to come from and cannot be formed on a screen.

● Reflection of light

When light rays strike a mirror or similar surface, they return at the same angle from the normal as the incident ray. This is called **reflection**.

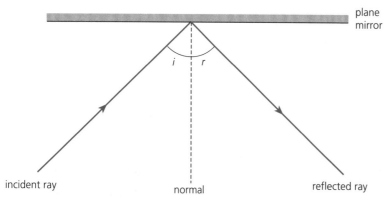

Figure 3.9 Reflection of light by a plane mirror

i = angle of incidence = angle between **normal** and incident ray
r = angle of reflection = angle between **normal** and reflected ray
angle of incidence = angle of reflection or $i = r$

● Sample question

Draw a diagram to show the path of a ray striking a plane mirror with an angle of incidence of 35°. Mark and label the incident ray, normal, reflected ray and angles of incidence and reflection. [4 marks]

Student's answer

The student's answer is shown by dashed lines in Figure 3.10. [2 marks]

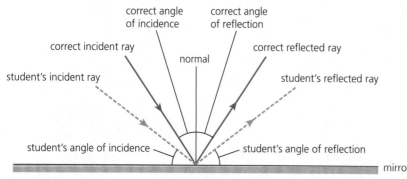

Figure 3.10

Cambridge IGCSE Physics Study and Revision Guide Second Edition © Mike Folland 2016

Examiner's comments

The student has measured the angle of incidence away from the mirror line, not away from the normal. The normal is correct, as is the reflected ray for the incident ray drawn. The angle of reflection is also incorrectly measured away from the mirror line.

Correct answer

The correct answer is shown by the solid lines in Figure 3.10.   [4 marks]

● Common error

✖ Finding the angles of incidence and reflection by measuring between the ray and the mirror.

● Formation of a virtual image by a plane mirror

A real image is one that can be produced on a screen and is formed by rays that actually pass through it.

   Figure 3.11 shows the formation of a virtual image by a plane mirror.

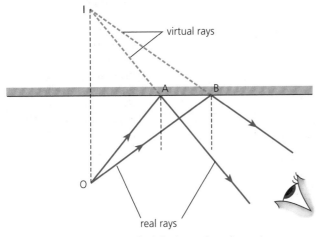

Figure 3.11 Construction to find the image in a plane mirror

The image of the object O is not formed on a screen. It is at point I where the rays appear to come from. The properties of an image in a plane mirror are:

● it is the same size as the object
● the line joining the object and the image is perpendicular to the mirror
● it is the same distance behind the mirror as the object is in front of the mirror
● it is laterally inverted

● it is virtual.

Examiner's tips

● You must be able to draw simple constructions.
● Hints for drawing a construction to show the position of the image of a point object in a plane mirror:

  ● Carefully measure the distance of the object from the mirror.

  ● Mark the image the same distance behind the mirror as the object is in front of the mirror. The object and image should be on a line at right angles to the mirror line (OI in Figure 3.11).

  ● Draw two lines from the image towards the eye; draw dotted lines behind the mirror where they represent virtual rays.

  ● Join up the two lines from the object to where the previous two lines cut the mirror line (A and B in Figure 3.11).

  ● Mark arrows on the real rays and label the diagram as necessary.

Cambridge IGCSE Physics Study and Revision Guide Second Edition © Mike Folland 2016

● Refraction of light

When a ray is travelling at an angle to a surface and enters a material where it travels slower, it changes direction *towards* the normal. This is called **refraction**. When a ray leaves this material, it is refracted *away from* the normal.

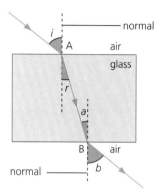

Figure 3.12 Refraction of a ray through a glass block

For refraction at point A, where the ray enters the glass:

- i = angle of incidence = angle between *normal* and incident ray
- r = angle of refraction = angle between *normal* and refracted ray
- the ray is refracted towards the normal, so the angle of refraction is less than the angle of incidence.

For refraction at point B, where the ray leaves the glass:

- a = angle of incidence = angle between *normal* and incident ray
- b = angle of refraction = angle between *normal* and refracted ray
- passing from glass to air, the ray is refracted away from the normal, so the angle of refraction is greater than the angle of incidence. When the block is parallel sided, the ray leaving is parallel to the ray entering.

Examiner's tips

- You will need to be able to describe an experiment to demonstrate the refraction of light.

- You should do this experiment in a darkened room. Direct a narrow beam of light at an angle to the side of a glass block placed on a large piece of paper. Mark on the paper the paths of the beams entering and leaving the block. By joining up the lines after removing the block, you can draw the path of the light as it travelled through the block. Figure 3.12 shows the paths of the rays in this experiment.

● Common error

✘ Finding the angles of incidence and refraction by measuring between the ray and the surface.

● Sample question

Copy and complete Figure 3.13 to show the path of the ray through the glass prism as it is refracted twice. Show *both* normals.   [4 marks]

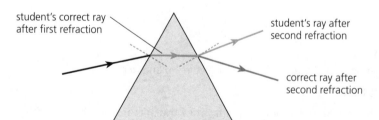

Figure 3.13 The black lines show the question.

Cambridge IGCSE Physics Study and Revision Guide Second Edition © Mike Folland 2016

Student's answer

The student's answer is shown by the orange lines in Figure 3.13.   [3 marks]

Examiner's comments

The student has correctly drawn the refracted ray within the prism and both normals. However, the student's second refraction, as the ray leaves the prism, is towards the normal. When a ray moves into a region where it travels faster, it is refracted away from the normal.

Correct answer

The correct rays are shown by the green line in Figure 3.13.   [4 marks]

● Refractive index

The amount of refraction is determined by the **refractive index** – the ratio of the speed of light in air to the speed of light in the material.

For example, to find the refractive index of glass, $n = \dfrac{\text{speed of light in air}}{\text{speed of light in glass}}$

● Sample question

Light travels at 3×10^8 m/s in air and at 2.25×10^8 m/s in water. Calculate:

1 the refractive index, n, of water

2 the angle of refraction for a ray approaching water with an angle of incidence of $55°$.   [4 marks]

Student's answer

1 $n = \dfrac{3 \times 10^8}{2.25 \times 10^8} = 1.33$ [2 marks]

2 $r = 34°$ [0 marks]

Examiner's comments

1 Correct answer with working.

2 The answer is only slightly inaccurate, but there is no working, which means the examiner has no way of knowing whether the student has made a small mistake or was completely wrong and simply close to the correct answer by chance. Examiners can only give credit for what they see.

Correct answer

1 $n = \dfrac{3 \times 10^8}{2.25 \times 10^8} = 1.33$ [2 marks]

2 $\sin r = \dfrac{\sin i}{n} = \dfrac{\sin 55}{1.33} = \dfrac{0.8192}{1.33} = 0.616$

$r = 38°$ to 2 s.f.   [2 marks]

Cambridge IGCSE Physics Study and Revision Guide Second Edition © Mike Folland 2016

Now try this

The answers are given on p. 118.

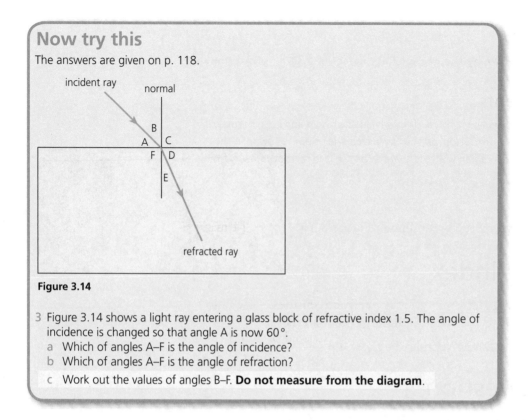

Figure 3.14

3 Figure 3.14 shows a light ray entering a glass block of refractive index 1.5. The angle of incidence is changed so that angle A is now 60°.
  a Which of angles A–F is the angle of incidence?
  b Which of angles A–F is the angle of refraction?
  c Work out the values of angles B–F. **Do not measure from the diagram**.

● Internal reflection and critical angle

Figure 3.15 shows a ray inside a block of glass or tank of water passing out into the air; some light is reflected internally and some is refracted away from the normal.

  i = angle of incidence = angle between *normal* and incident ray

  r = angle of internal reflection = angle between *normal* and internally reflected ray

  R = angle of refraction = angle between *normal* and refracted ray

The law of reflection still applies, so $i = r$.

The greater the angle of incidence, the more energy goes into the internally reflected ray, which becomes brighter. The greatest angle of incidence when refraction can still occur is called the **critical angle**; in this case (shown in Figure 3.16), the angle of refraction is 90° and the refracted ray travels along the surface.

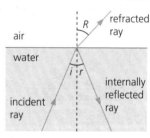

Figure 3.15 Rays at a water–air boundary

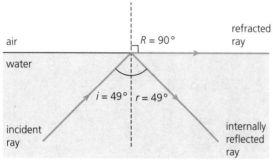

Figure 3.16 Angle of incidence is the same as the critical angle

In this case, sin R = sin 90 = 1 and the refractive index is $1/n$ because the light is passing from water into light. The refraction equation becomes: $\frac{1}{n} = \frac{\sin c}{1}$ or $n = \frac{1}{\sin c}$.

You should know and be able to use the equation $n = \frac{1}{\sin c}$.

If the angle of incidence is greater than the critical angle, there is no refracted ray and all of the energy is in the bright internally reflected ray. This is called **total internal reflection** (Figure 3.17).

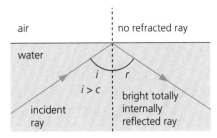

Figure 3.17 Total internal reflection

● Sample question

A ray of light is in water of refractive index 1.33. The ray approaches the interface with air at an angle of incidence of 52°. Carry out a suitable calculation and state what happens to the ray after striking the interface. [4 marks]

Student's answer

$n = \frac{\sin i}{\sin r}$

$\sin r = \frac{\sin 52°}{1.33} = 0.592$

$r = 36°$ [1 mark]

Examiner's comments

The student failed to make a final statement.
The student used the wrong value for refractive index.

Correct answer

For light passing from water to air, the correct value is $n = \frac{1}{1.33}$

$n = \frac{\sin i}{\sin r}$

$\frac{1}{1.33} = \frac{\sin i}{\sin r}$

$\sin r = 1.33 \times \sin 52° = 1.05$

It is impossible to have a sin value greater than 1, which indicates that the critical angle must have been exceeded and total internal reflection occurs. This would be a completely correct answer.

If a student happens to recognise that this is possibly the case, there is an alternative approach.

Cambridge IGCSE Physics Study and Revision Guide Second Edition © Mike Folland 2016

Alternative approach:

The angle of incidence is close to the critical angle, so check that first.

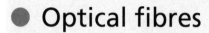

$$\sin c = \frac{1}{1.33}$$

$c = 48.75°$ [3 marks]

The angle of incidence is confirmed as greater than the critical angle, so total internal reflection occurs and the ray is reflected back in the water with an angle of reflection of $52°$.   [1 mark]

● Optical fibres

Each time the light strikes the wall of the optical fibre, the angle of incidence is greater than the critical angle and so total internal reflection occurs. There is very little loss of energy. The light can be considered 'trapped' in the optical fibre and can travel long distances, even if the fibre is bent, in order to carry information or illuminate and view inaccessible places.

Figure 3.18 Light travels through optical fibre by total internal reflection

You should be able to describe and explain the use of optical fibres in medicine and telecommunication.

Now try this

The answers are given on p. 118.

4 a Describe the action of optical fibres in a medical application.
  b Describe the action of optical fibres in a telecommunication application.

Cambridge IGCSE Physics Study and Revision Guide Second Edition © Mike Folland 2016

Lenses

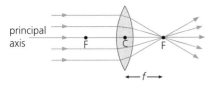

Figure 3.19 Action of a converging lens on a parallel beam of light

All rays of light parallel to the principal axis are refracted by the lens to pass through the **principal focus**, F. The distance between F and the optical centre, C, is called the focal length, *f*.

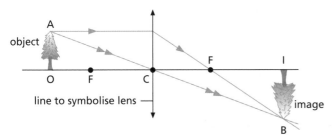

Figure 3.20 Ray diagram for a converging lens

If the object is placed more than one focal length behind the lens, the image will always be real and inverted. Depending on where the object is placed, the image may be magnified, the same size as the object or diminished.

● Common error

✖ Inaccurate drawing work leading to very poor answers.
✔ Because rays intersect at small angles, errors can be easily magnified. It is essential to work carefully with a sharpened pencil; distances should be measured to within 0.5 mm or less and lines drawn *exactly* through points.

Cambridge IGCSE Physics Study and Revision Guide Second Edition © Mike Folland 2016

Now try this

The answers are given on p. 119.

5 An object of height 1.5 cm is placed 4.5 cm from a thin converging lens of focal length 3 cm. Draw a ray diagram to find:
   a   the nature of the image
   b   the size of the image
   c   the position of the image.

Examiner's tip

Know and be able to use the following descriptions of an image when it is compared with the object: enlarged/the same size/diminished/upright/inverted.

● Formation of a virtual image by a converging lens

If the object is placed closer to the lens than the principal focus, the rays leaving the lens do not converge to form a real image.

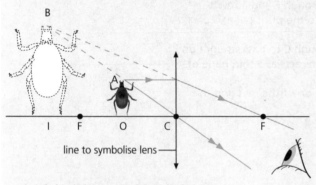

line to symbolise lens

Image is behind object, virtual, upright, larger

Figure 3.21 Ray diagram for a converging lens used as a magnifying glass

Magnifying glass

Figure 3.21 shows how a converging lens can be used as a magnifying glass. The object is placed less than one focal length behind the lens. No real image is formed, but the eye sees the rays diverging from the magnified virtual image. O is the object and I the image.

Now try this

The answers are given on p. 119.

6 An object of height 1.5 cm is placed 2 cm from a thin converging lens of focal length 3 cm. Draw a ray diagram to show two rays from the object passing through the lens.
   a   What can you say about the two rays as they leave the lens?
   b   With dotted lines, draw your rays back behind the object. Find:
      (i)   the nature of the image
      (ii) the size of the image
      (iii) the position of the image.

Examiner's tip

Know and be able to use the terms 'real' and 'virtual' when describing an image.

Cambridge IGCSE Physics Study and Revision Guide Second Edition © Mike Folland 2016

Now try this

The answers are given on p. 119.

7 A surveyor's telescope used to observe a measuring pole with a scale produces an upside down image compared with the object.
a Suggest and explain a helpful way to have the numbers of the scale printed.
b The telescope's image can be seen only in the eyepiece and cannot be projected on a screen.
Underline two items from the following list that describe the image (for extended paper, supply three items):

enlarged
same size
diminished
upright
inverted
real
virtual

● Dispersion of light

White light is made up of seven colours. Each colour is refracted by a different amount in glass. If a beam of white light falls on a glass prism, it is dispersed into a **spectrum** of the seven colours (Figure 3.22).

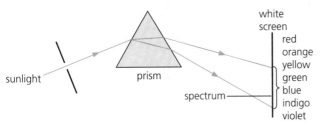

Figure 3.22 Forming a spectrum with a prism

Monochromatic light

Light of a single frequency or wavelength is described as monochromatic, which means of a single colour.

Cambridge IGCSE Physics Study and Revision Guide Second Edition © Mike Folland 2016

Electromagnetic radiation

● Electromagnetic waves

All types of waves that make up the electromagnetic spectrum have properties in common:

- They can travel through a vacuum at the same high speed, which is much faster than other types of waves that travel through a material.
- They show the normal wave properties of reflection, refraction and diffraction.
- They are transverse waves.
- They travel owing to moving electric and magnetic fields.

The speed of electromagnetic waves in a vacuum is 3×10^8 m/s. The speed is approximately the same in air.

Now try this

The answers are given on p. 119.

8 a Calculate the frequency of green light of wavelength 5.5×10^{-7} m.

b Use your knowledge of the electromagnetic spectrum to deduce and choose from the following list a possible wavelength for red light:

5.5×10^{-10} m

4.1×10^{-7} m

5.5×10^{-7} m

7.1×10^{-7} m

5.5×10^{-4} m

Examiner's tip

Remember that all electromagnetic waves have the same speed in a vacuum and the wave equation, $v = f\lambda$, applies, so the higher the wavelength the lower the frequency.

The Sun and other stars give off a wide range of types of electromagnetic waves, which travel through space to Earth. Much of this radiation is stopped by Earth's atmosphere and can be detected only by satellites in orbit outside the atmosphere.

The types of electromagnetic waves, in order of increasing wavelength, are described below.

Gamma rays are produced by radioactive substances. They are very dangerous to living matter. They are used to kill cancer cells and dangerous bacteria.

X-rays are produced in high-voltage X-ray tubes. They are absorbed differently by different types of matter. They can produce shadow pictures of inside the human body, which are invaluable for medical diagnosis. X-rays can penetrate inaccessible solid structures. They are used by security machines at airports and other travel hubs to scan luggage for dangerous hidden objects. X-rays are dangerous to living matter as they can kill cells and cause cancers. Lead shielding must be used to protect people from exposure, especially those who work regularly with X-rays.

Cambridge IGCSE Physics Study and Revision Guide Second Edition © Mike Folland 2016

Ultraviolet (UV) radiation is produced by the Sun, special UV tubes and welding arcs. The radiation can cause sunburn and skin cancer; it also produces vitamins in the skin and causes certain substances to fluoresce. This fluorescence can reveal markings that are invisible in normal light.

Visible light is a very narrow range of wavelengths that can be seen by the human eye as the colours of the visible spectrum from violet to red.

Infrared (IR) radiation is produced by hot objects and transfers thermal energy to cooler objects. Hot objects below about $500\,°C$ produce IR radiation only; above this temperature, visible light is also radiated. Night-vision goggles detect the IR radiation given off by warm objects.

Microwaves are used for telecommunication, radar and cooking. Satellite and mobile phone telecommunication, as well as Bluetooth and Wi-Fi systems, use microwaves. Microwaves can damage living cells and, as they travel through matter, cause internal burns. Parts of the ears and eyes, in particular, are easily damaged by microwaves, so great care is needed to ensure that the doors of microwave ovens are always closed when in use and that excessive mobile phone use is avoided. Personnel servicing military aircraft in an operational situation wear protective suits to reflect the microwaves emitted by the high-powered radar in the aircraft.

Radio waves come in a wide range of wavelengths that can be split into the following groups:

- Very high frequency (VHF) and ultra high frequency (UHF), of shorter wavelengths, are used for television and radio.
- Traditional radio waves (short, medium and long wave) have the longest wavelengths.

● Common error

✖ Using the expression 'ultraviolet light'.
✔ The correct expression is 'ultraviolet radiation' – UV radiation is not part of the visible spectrum, so must *not* be called 'light'. This misconception might occur because normally UV lamps give off blue and violet light as well as UV radiation.

● Sample question

1 Which of the following types of waves is *not* an electromagnetic wave: light, X-rays, radio, sound? [1 mark]

2 State one property that this wave has in common with electromagnetic waves. [1 mark]

3 State one property of this wave that is different from electromagnetic waves. [1 mark]

Student's answer

1 Sound. [1 mark]
2 Wavelength. [0 marks]
3 Reflection. [0 marks]

Cambridge IGCSE Physics Study and Revision Guide Second Edition © Mike Folland 2016

Examiner's comments

1 *Correct answer.*

2 *Incorrect answer – although both types of waves do have a wavelength, its value can vary over a wide range, so it is unlikely to be the same for each type of wave.*

3 *Incorrect answer.*

Correct answers

1 Sound.   [1 mark]

2 It can be reflected *or* It can be refracted *or* It can be diffracted.   [1 mark]

3 Sound waves are longitudinal but electromagnetic waves
are transverse.   [1 mark]

> ### Now try this
> The answers are given on p. 119.
>
> 9 An observatory receives X-rays and gamma rays from a star.
>   a Which type of radiation has the higher wavelength?
>   b Which type of radiation has the higher frequency?
>   c Light from the star takes 4 years to reach the observatory. Do the X-rays take less, more or the same time to reach the observatory?
>
>   d Work out the distance from the observatory to the star in kilometres.

Sound

Sound waves are **longitudinal** waves that are produced by a vibrating source, which causes a material to vibrate. A material or medium is required to transmit sound waves.

Although normally observed in air, sound waves can travel through liquids and solids, e.g. sea creatures communicate by sound waves travelling through water.

The healthy human ear can hear sound in air in the frequency range of 20 Hz to 20 000 Hz (20 kHz). This is called the **audible range**. In practice, only people with very good hearing can hear throughout this range. With ageing, this range is reduced and hearing tests check only between 250 Hz and 8 kHz.

Sound of a higher frequency than the audible range is called **ultrasound**.

The greater the amplitude of sound waves, the louder the sound.

The greater the frequency of sound waves, the higher the pitch.

Sound waves can be reflected, especially from large, hard, flat surfaces. The reflected sound is called an **echo**.

As sound travels through a material, compressions and rarefactions occur (see Figure 3.2 on p. 48). Compressions are regions where particles of material are closer together. Regions of material are rarefied where the particles move further apart (at rarefactions).

(see Figure 3.2 on p. 48)

Examiner's tips

- You must be able to describe an experiment to find the speed of sound in air. (See the sample question.)

- The speed of sound in air at normal temperatures is about 340 m/s. Sound travels faster in water, at about 1500 m/s. The speed of sound in solids is very high and in steel is 5000 m/s. Extended candidates should know these typical values.

● Sample question

A student stands 100 m from a large building and claps her hands regularly at a rate of 16 claps every 10 s. She hears each clap coincide exactly with the echo from the clap before. Work out the speed of sound in air.   [4 marks]

Student's answer

Time between claps $= \frac{16}{10} = 1.6$ s   [0 marks]

Distance travelled $= 2 \times 100 = 200$ m   [1 mark]

Speed $= \frac{200}{1.6} = 125$ m/s   [1 mark]

Examiner's comments

The student calculated the number of claps per second instead of the time from one clap to the next. No further error was made in calculating the speed, so the student gained the last two marks.

Cambridge IGCSE Physics Study and Revision Guide Second Edition © Mike Folland 2016

Correct answer

Time between claps $= \dfrac{10}{16} = 0.625$ s   [2 marks]

Distance travelled $= 2 \times 100 = 200$ m   [1 mark]

Speed $= \dfrac{200}{0.625} = 320$ m/s   [1 mark]

● Common errors

✖ In problems involving echoes, taking the distance from the observer to the reflecting surface as the distance travelled by the sound.

✔ When the source and observer are at the same place, the sound travels twice the distance between the observer and the reflecting surface.

✖ Doubling the distance when no echo or reflection is involved.

● Sample question

A railway worker gives a length of rail a test blow with a hammer, striking the end of the rail in the direction of its length. A sound of frequency 10 kHz travels along the rail. Use your knowledge of a typical speed of sound in a solid to estimate the wavelength of the sound.   [4 marks]

Student's answer

Estimated speed of sound $= 3000$ m/s   [2 marks]

Wavelength, $\lambda = \dfrac{v}{f} = \dfrac{3000}{10\,000} = 0.3$ m   [2 marks]

Examiner's comments

The estimate for the speed of sound is on the low side, as sound travels faster in steel than in less rigid solids. Full marks are awarded, as only a typical speed was asked for. The rest of the answer is correct.

Correct answer

Estimated speed of sound $= 5000$ m/s   [2 marks]

Wavelength, $\lambda = \dfrac{v}{f} = \dfrac{5000}{10000} = 0.5$ m   [2 marks]

● Sample question

1 A machine uses sound waves of frequency 3 MHz to form images within the human body.

Underline one of the following as the best description of these waves:

long wavelength

hypersound

polarised sound

supersonic

ultrasound   [1 mark]

Cambridge IGCSE Physics Study and Revision Guide Second Edition © Mike Folland 2016

2 Dolphins emit sound waves of 95 kHz. State and explain if the best description from question 1 applies to these waves. [2 marks]

3 At a concert attended by people of all ages including children and old people, a sound of frequency 19 kHz is produced. Comment on how this would be heard by the audience. [1 mark]

Student's answer

1 *polarised sound* [0 marks]

2 *Sound waves of 95 kHz are ultrasound because they travel faster than normal sound waves.* [1 mark]

3 *Everyone would hear it because it is in the audible range.* [1 mark]

Examiner's comments

1 Incorrect response.

2 Ultrasound is correct but the reason is incorrect.

3 The answer is on the right lines but is incomplete because in such an audience it is unlikely that everyone would have completely healthy ears.

Correct answers

1 Ultrasound. [1 mark]

2 Ultrasound, because the frequency is higher than the audible range. [2 marks]

3 People with healthy hearing would hear the sound of 19 kHz. It is right at the top of the audible range, so people with any loss of high-frequency hearing would not hear it. [1 mark]

> ### Now try this
> The answer is given on p. 119.
>
> 10 A research ship uses an echo-sounder to locate a shoal of fish. It receives back an echo 37.1 ms after a sound is transmitted. Work out the depth of the shoal below the ship. (Speed of sound in water = 1500 m/s. This figure would not be given in the extended paper.)

Cambridge IGCSE Physics Study and Revision Guide Second Edition © Mike Folland 2016

TOPIC 4 — Electricity and magnetism

Main points to revise

- the properties of magnets
- how to describe the magnetic field due to electric currents
- the principle of electromagnetic induction and how it is applied in an a.c. generator and a transformer
- the basic hazards of electricity
- how a magnetic field can cause a force on a current-carrying conductor and how to relate this force to a d.c. electric motor

- how to solve problems involving current, voltage and resistance in circuits
- knowledge of the use and symbols of circuit compoments
- how to describe the action of logic gates on their own and in circuits
- how to describe the deflection of beams of charged particles in magnetic fields.

● Key terms

Term	Definition
a.c.	Alternating current is current that flows in alternating directions at different stages in the cycle
d.c.	Direct current is current that flows in only one direction
Magnetic field	A region where a magnet experiences a force
Electric charge	An excess or lack of electrons measured in coulombs
Electric field	A region where an electric charge experiences a force
Current	The rate of flow of charge measured in amps
e.m.f.	Electromotive force is the source of electrical energy in a circuit measured in volts
p.d.	Potential difference is the voltage between two points in a circuit
In series	Circuit components arranged so that current flows from one component to another
In parallel	Circuit components arranged so that current has alternative routes to flow through any one component
Resistance	$\text{resistance} = \dfrac{\text{p.d.}}{\text{current}}$ measured in ohms
Volts	Joules per coulomb
Amps	Coulombs per second
LDR	The resistance of a light-dependent resistor (LDR) depends on the intensity of light falling on it
Thermistor	The resistance of a thermistor depends on its temperature
Analogue	In an analogue circuit, the voltages can vary continuously to take any value in the available range.
Digital	In a digital circuit, voltage can only be in one of two states, either zero or maximum voltage.

Cambridge IGCSE Physics Study and Revision Guide Second Edition © Mike Folland 2016

Magnetism

A magnetic field is a region where a magnetic material experiences a force.

Magnetic materials are chiefly the ferrous metals iron and steel and their alloys. Cobalt, nickel and certain alloys are also magnetic materials.

A magnetic field can be produced by a permanent magnet or a wire carrying an electric current. A magnetic field also exists around the Earth due to convection currents in its core.

The direction of a magnetic field is the direction of the force on the N pole of a magnet at that point.

Examiner's tips

- You must be able to draw the pattern of magnetic field lines around a bar magnet and describe an experiment to identify this pattern, including the directions.

- This experiment can be done with either a plotting compass or iron filings.

- If iron filings are used, a plotting compass or suspended magnet is needed to confirm the direction.

● Drawing magnetic fields

Plotting compass method: The magnet is placed on a sheet of paper and a small plotting compass is placed near one pole. Mark dots on the paper at the positions of the ends of the compass needle. The compass is moved along so that the end that was over the first dot is now over the second dot. The other end is marked on the paper as the third dot. Continue this process until the other pole of the magnet is reached. Joining the dots with a smooth line shows the field, with the direction being given by the compass arrow. Repeat for further lines starting at different points.

Iron filings method: Iron filings are sprinkled on a piece of paper placed over a magnet. When the paper is tapped gently, the filings will be seen to line up with the fields lines. The field pattern can then be drawn along these lines on the paper.

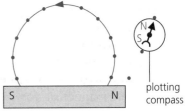

Figure 4.1 Plotting field lines

● Sample question

Describe how to plot the magnetic field, including its direction, around a bar magnet using iron filings.　　　　　　　　　　　　　　　　[5 marks]

Student's answer

The iron filings should be spread around the magnet and the pattern drawn.
Use a plotting compass to find the direction.　　　　　　　　[1 mark]

Examiner's comments

The student basically knows the right experiment but the description is very vague and lacking in essential detail.

Correct answer

To gain full marks, the student should have given the following answer:

Place a piece of paper on top of the bar magnet. [1 mark] Sprinkle iron filings thinly and evenly over the paper. [1 mark] Give the paper a gentle tap. [1 mark] Draw the field pattern on the paper along the lines of the filings. [1 mark] Place a plotting compass on top of a field line and draw a direction arrow in the direction of the N end of the needle. Repeat to establish the directions of the whole field pattern. [1 mark]

Cambridge IGCSE Physics Study and Revision Guide Second Edition © Mike Folland 2016

● Magnets

One end of a magnet suspended in the Earth's magnetic field will swing around to point towards the Earth's magnetic north pole. This is called the N pole of the magnet. The other end of the magnet is the S pole.

If two magnets are close together, poles N and N will repel, poles S and S will repel but poles N and S will attract.

> ## Now try this
>
> The answers are given on p. 119.
>
> 1 A bar magnet is suspended from its mid-point by a thread.
>   a Which pole will swing towards the Earth's magnetic north pole?
>   b Another bar magnet is brought close to the pole of the suspended magnet that is furthest away from the Earth's magnetic north pole and does *not* cause the suspended magnet to swing around.
>   Which pole of this magnet is closest to the suspended magnet?

Examiner's tips
- Unlike poles attract.
- Like poles repel.
- The Earth's magnetic north pole attracts the N pole of a magnet.
- Magnetic forces are interactions between magnetic fields.

● Magnetisation

If a piece of iron or steel is placed in a magnetic field it too will become magnetised. This is called induced magnetism.

Steel is hard to magnetise by induction but once magnetised keeps its magnetism. Materials that are hard to magnetise and demagnetise are called magnetically hard materials. Permanent magnets, which need to maintain a high level of magnetism for a long time, are made from these materials.

Uses of permanent magnets include loudspeakers, electrical meters and small electric motors such as those used in domestic appliances.

Generally, iron is easily magnetised but loses its induced magnetism when removed from the magnetic field. Materials that easily magnetise and demagnetise are called magnetically soft materials.

Magnets that need to be switched on and off readily or have their fields reversed are made from soft iron. Their uses include electromagnets, large-scale motors and generators, relays and transformers.

● Common errors

✖ Thinking that magnets attract all metals.
✔ Magnets can attract only ferrous metals, cobalt, nickel and certain alloys.

Materials can be magnetised by placing them in a strong magnetic field. This is best done in the field produced inside a long coil, or solenoid, by a high direct current (d.c.) flowing through it.

Other methods of magnetising a steel bar are striking it a sharp blow with a hammer in a strong magnetic field or 'stroking' another magnet along the length of the bar repeatedly in the same direction.

Cambridge IGCSE Physics Study and Revision Guide Second Edition © Mike Folland 2016

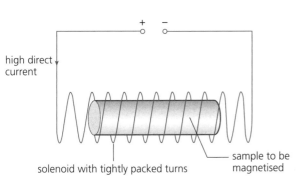

Figure 4.2 Magnetising in a solenoid

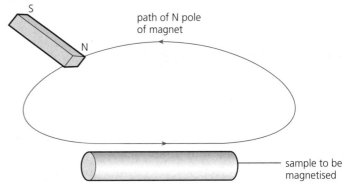

Figure 4.3 Magnetising by stroking

Simple methods of demagnetising a material are by heating until red hot or hammering with the magnet lying in an east–west direction.

A material can also be demagnetised electrically. The magnet is placed in a solenoid supplied with an alternating current (a.c.). The current is gradually reduced to zero. During this process, the material is magnetised in opposite directions every half cycle of the a.c. but less and less strongly to nothing as the current falls to zero.

● Sample question

One method of demagnetising a bar magnet is to place it in a solenoid with a high alternating current and then slowly withdraw the magnet from the solenoid. Explain how this demagnetises the magnet.   [4 marks]

Student's answer

The magnet gets tired of the alternating field in the solenoid and loses magnetism.   [1 mark]

Examiner's comments

The student has correctly recognised that there is an alternating field inside the solenoid but fails to understand the principle of magnetisation.

Correct answer

The alternating current causes the direction of the field of the solenoid to alternate. [1 mark] The magnet when inside the solenoid is repeatedly magnetised in alternating directions by the strong alternating field. [1 mark] Outside the solenoid, the field weakens as distance increases. As each part of the magnet is withdrawn, it is magnetised in alternating directions less and less strongly [1 mark] until no magnetism remains. [1 mark]

Now try this

The answers are given on p. 119.

2 Non-magnetised bars of hard steel, soft iron and copper are placed in and then removed from a coil carrying a very high d.c. current. The bars are then tested by seeing how many paper clips they can pick up. Which bar picks up:
  a   no paper clips
  b   a few paper clips
  c   many paper clips?

Static electricity

Electric charge

An **electron** is a subatomic particle with a negative electric charge.

Certain materials can lose or gain electrons.

Atoms, and objects composed of atoms, are normally electrically neutral. The number of electrons (negative charges) balances the number of positive charges in each atom. If an object gains extra electrons, it becomes negatively charged. If it loses electrons, it becomes positively charged.

If two similarly charged objects are close together (+ and +, or − and −) they will repel, but unlike charges (+ and −) will attract.

The charge on an electron is the smallest possible quantity of charge. Charge is measured in coulombs (C). The charge on one electron is 1.6×10^{-19} coulombs, but you do not need to remember that number.

Examiner's tips
- Unlike charges attract.
- Like charges repel.

Common errors

✖ There is a force of attraction only when both objects are charged.
✔ Charged objects attract both uncharged and charged objects because the charged object can induce a charge on the uncharged object.
✖ Positive charges can move.
✔ Objects become positively charged because electrons (negative charges) move away.

Positive and negative charge

When certain materials (e.g. polythene) are rubbed with a cloth, electrons are moved from the cloth on to the polythene. The polythene gains electrons and becomes **negatively** charged.

When certain other materials (e.g. cellulose acetate) are rubbed with a cloth, electrons are moved from the cellulose acetate on to the cloth. The cellulose acetate loses electrons and becomes **positively** charged.

The fact that an object is electrically charged can be detected as shown in Figure 4.4.

Conductors and insulators

Electrical conductors are materials in which electrons can move freely from atom to atom, so electric charge can flow readily. All metals and some forms of carbon are conductors.

Electrical insulators are materials in which electrons are firmly held in their atoms and cannot move, so electric charge cannot flow. Most plastics are good insulators.

In Figure 4.4, two polythene rods are rubbed with a cloth to become negatively charged. One rod is suspended freely on a thread. The other rod is brought close to one end of the suspended rod, which rotates away owing to repulsion between like charges.

Similarly, if a charged cellulose acetate rod (positively charged) is brought close to one end of the suspended,

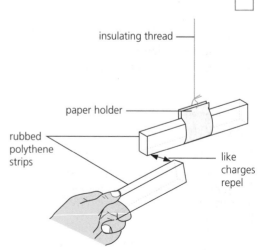

insulating thread

paper holder

rubbed polythene strips

like charges repel

Figure 4.4 Investigating charges

Cambridge IGCSE Physics Study and Revision Guide Second Edition © Mike Folland 2016

charged polythene rod, the latter will rotate closer owing to attraction between unlike charges.

Another test for electric charge is to use a **gold-leaf electroscope**.

● Sample question

An inkjet printer produces a stream of very small droplets from a nozzle. The droplets are given a negative electric charge and then pass between two plates with positive and negative charge as shown in Figure 4.5.

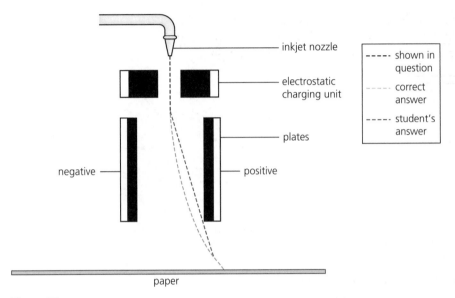

Figure 4.5

1 State the type of field in the region between the charged plates.   [1 mark]

2 State and explain the force acting on the droplets in this region.   [2 marks]

3 Extend the dashed line to complete the path of the droplets.
                                                    [2 marks] or [3 marks]

Student's answer

1 There is an electric field in this region.   [1 mark]
2 There is a force to the right on the ink droplets.   [1 mark]
3 The student's answer is shown by the red dashed line in Figure 4.5.   [2 marks]

Examiner's comments

1 Correct answer.

2 The student correctly stated that the force is to the right but failed to explain it.

3 Acceptable answer for core students.

The path between the plates must be curved, as the droplets are accelerated to the right by the electric force.

Correct answer

1 There is an electric field in this region.   [1 mark]

2 There is a force to the right on the ink droplets, which are attracted
   to the positive plate.   [2 marks]

Cambridge IGCSE Physics Study and Revision Guide Second Edition © Mike Folland 2016

3 An acceptable answer for core candidates would be either the red or the blue dashed lines in Figure 4.5.   [2 marks]

The correct answer is shown by the blue dashed line in Figure 4.5.   [3 marks]

● Electric field

An **electric field** is a region where an electric charge experiences a force.

The field can be represented by lines of force in the direction that a positive charge would move if placed at a point in the field (Figures 4.7 and 4.8).

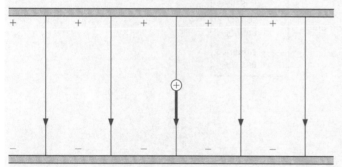

Figure 4.6 Uniform field between parallel conducting plates with opposite charge

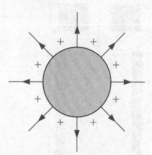

Figure 4.7 Radial field around a positively charged conducting sphere

A conducting object can be charged by induction as shown in the example in Figure 4.8, with two conducting spheres. The neutral, uncharged metal spheres are in contact and a charged rod is brought close. This repels electrons, charging the right side of sphere Y negatively and the left side of sphere X positively (Figure 4.8a). Keeping the charged rod close to sphere X, sphere Y is moved away (Figure 4.8b). When the charged strip is removed, the spheres have been charged (Figure 4.8c).

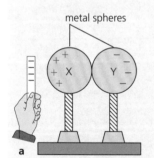
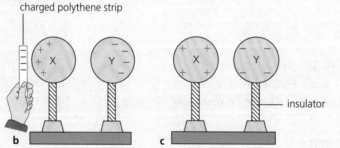

Figure 4.8 Electrostatic induction

Now try this

The answers are given on p. 119.

3 An insulating piece of plastic is given a large electric charge and placed near a thin stream of water flowing vertically down from a nozzle. Describe:
   a the effect on the charges within the stream of water
   b the forces exerted on these charges
   c the position of the stream of water.

Electric current

Electric current is a flow of charge. Electric current flows from the '+' terminal of a battery or power supply through the circuit to the '−' terminal. In metals, current is due to a flow of electrons.

Electric current is measured in **amperes**, usually abbreviated to amps (symbol A), by an ammeter, which must be connected **in series**.

One ampere is a flow of one **coulomb** (1 C) of charge in 1 s.
To work out the current, you need to use the equation $I = \frac{Q}{t}$, where I = current, Q = charge and t = time.

Figure 4.9 shows an ammeter in series with component X.

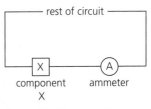

Figure 4.9 Measuring the current through component X

● Common error

✖ Putting an ammeter **in parallel** with the part of the circuit where current is to be measured.

✔ The current to be measured must flow through the ammeter, which needs to be connected **in series**.

In a circuit, we can consider that positive charge is repelled by the positive terminal and attracted by the negative terminal, so charge and current flow from positive to negative. This is called the 'conventional current'. In reality, negatively charged electrons flow in the opposite direction, from negative to positive.

● Sample question

A charge of 35 C flows around a circuit in 14 s.

1 Work out the current flowing.

2 The charge on each electron is 1.6×10^{-19} C.

Work out the number of electrons flowing around the circuit in this time. [4 marks]

Student's answer

1 $I = \frac{Q}{t} = \frac{35}{14} = 2.5 \text{ A}$ [2 marks]

2 Number of electrons $= \frac{35}{1.6 \times 10^{-19}} = 2.19 \times 10^{19}$ [1 mark]

Examiner's comments

1 Correct answer.

2 The student started the calculation correctly but made a mistake in the calculation of the powers of 10.

Correct answer

1 $I = \frac{Q}{t} = \frac{35}{14} = 2.5 \text{ A}$ [2 marks]

2 Number of electrons $= \frac{35}{1.6 \times 10^{-19}} = 2.2 \times 10^{20}$ to 2 s.f.
(2.189×10^{20}) [2 marks]

Electromotive force and potential difference

Electromotive force (e.m.f.) is the measure of the electrical energy transferred in a whole circuit. It is measured in volts. e.m.f. (in volts) is the amount of energy (in joules) given to each coulomb of charge as it passes around the circuit. 1 V is equivalent to 1 J/C.

Potential difference (p.d.) across a component is the measure of the electrical energy transferred in the component. It is measured in volts.

It is the amount of energy (in joules) given to each coulomb of charge as it passes through the component.

p.d. is measured by a voltmeter, which must be connected between the two points, **in parallel** with any circuit elements between the points.

Figure 4.10 shows a voltmeter in parallel with component X.

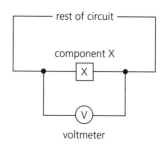

Figure 4.10 Measuring the p.d. across component X

● Common error

✖ Putting a voltmeter **in series** with the part of the circuit where p.d. is to be measured.

✔ The voltmeter must be connected **in parallel**.

Now try this

The answers are given on p. 119.

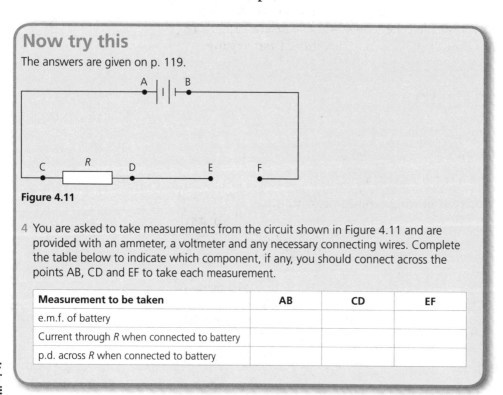

Figure 4.11

4 You are asked to take measurements from the circuit shown in Figure 4.11 and are provided with an ammeter, a voltmeter and any necessary connecting wires. Complete the table below to indicate which component, if any, you should connect across the points AB, CD and EF to take each measurement.

Measurement to be taken	AB	CD	EF
e.m.f. of battery			
Current through R when connected to battery			
p.d. across R when connected to battery			

Cambridge IGCSE Physics Study and Revision Guide Second Edition © Mike Folland 2016

Resistance

For a given p.d., the greater the resistance, the smaller the current.
  For a given resistance, the greater the p.d., the greater the current.
  You should know and be able to use the equation $R = \dfrac{V}{I}$, where
R = resistance in ohms (Ω), V = p.d. and I = current.

This equation is called Ohm's law. It is obeyed by wires and resistors at constant temperature, which are called ohmic resistors. The temperature of a filament lamp increases considerably with increasing current, as does its resistance.

● Experiment to determine resistance of unknown conductor Y

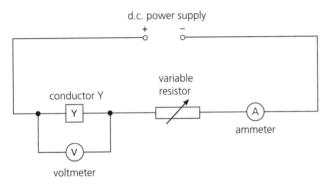

Figure 4.12

1 Connect the circuit as shown in Figure 4.12.

2 Ensure that the ammeter is **in series** with the power supply and conductor Y (so that the current flows *through* the ammeter).

3 Ensure that the voltmeter is **in parallel** with conductor Y (so that the voltmeter measures the p.d. *between* the ends of the conductor).

4 Record the values of the p.d. in volts and the current in amps.

5 Change the setting of the variable resistor and record at least five more pairs of values.

6 Work out the value of the resistance using the equation $R = \dfrac{V}{I}$ for each pair of readings.

7 Work out the average value of R, or use a graphical method.

● Sample question

A student carries out an experiment to find the resistance of a wire. He varies the supply voltage and takes measurements using an ammeter and a voltmeter. The table below shows his first three readings.

Reading	1	2	3	4
Voltage/V	1.10	2.10	2.95	1.75
Current/A	0.20	0.35	0.50	?
Resistance/Ω	5.50	6.0	?	?

Make sure you show your working for all parts of this question.

Cambridge IGCSE Physics Study and Revision Guide Second Edition © Mike Folland 2016

1 Work out the resistance for the third pair of readings. [1 mark]

2 Work out the average value of resistance. [1 mark]

3 Using this average value of resistance, work out the current the student can expect when he takes the fourth reading. [2 marks]

Student's answer

1 Reading 3: $R = \dfrac{V}{I} = \dfrac{2.95}{0.50} = 5.9\ \Omega$ [1 mark]

2 Average $V = \dfrac{1.10 + 2.10 + 2.95}{3} = \dfrac{6.15}{3} = 2.05\ V$

   Average $I = \dfrac{(0.20 + 0.35 + 0.50)}{3} = \dfrac{1.05}{3} = 0.35\ A$

   Average $R = \dfrac{2.05}{0.35} = 5.86\ \Omega$ [0 marks]

3 Reading 4: $I = VR = 1.75 \times 5.86 = 10.26\ A$ [0 marks]

Examiner's comments

1 Correct answer.

2 The student should have taken the average of the three values of resistance from question 1. It is simply good fortune that his answer is close to the correct answer.

3 The student has rearranged the equation $V = IR$ incorrectly. Observation and comparison with the first three readings show that the calculated value is unlikely.

Correct answer

1 Reading 3: $R = \dfrac{V}{I} = \dfrac{2.95}{0.50} = 5.9\ \Omega$ [1 mark]

2 Average $R = \dfrac{5.50 + 6.00 + 5.90}{3} = \dfrac{17.40}{3} = 5.8\ \Omega$ [1 mark]

3 Reading 4: $I = \dfrac{V}{R} = \dfrac{1.75}{5.80} = 0.30\ A$ to 2 s.f. (0.3017 A) [2 marks]

● The resistance of a wire

The longer the wire, the greater the resistance.
The thicker the wire, the smaller the resistance.

> Resistance is proportional to length of the wire.
>
> Resistance is proportional to $\dfrac{1}{\text{cross-sectional area of the wire}}$.

Examiner's tips

- Know and be able to use the above two relationships for resistance.

- Many students may find the easiest way to do this is to remember the following equation (even though it is not required by the syllabus):

$$\text{resistance} = \dfrac{\text{constant} \times \text{length}}{\text{cross-sectional area}}.$$

Cambridge IGCSE Physics Study and Revision Guide Second Edition © Mike Folland 2016

● Sample question

Sample A is a length of wire of given material.

1 Copy and complete the table below for the resistance of three more samples of wire of the same material. Choose from the following words: greater, less, same.

Sample	B	C	D
Length compared with A	×2	Same	×2
Diameter compared with A	Same	$\frac{1}{2}$	$\frac{1}{2}$
Resistance compared with A			

[3 marks]

2 Add numerical values to your entries in the table to show the magnitude of resistance compared with sample A. [3 marks]

Student's answer

Sample	B	C	D
Length compared with A	×2	Same	×2
Diameter compared with A	Same	$\frac{1}{2}$	$\frac{1}{2}$
Resistance compared with A	Greater ×2	Greater ×2	Greater ×4

[3 marks]

[2 marks]

Examiner's comments

1 Correct answers – all three samples have greater resistance than sample A.

2 The resistance of sample B will be 2× greater – the student's answer to B is correct. Resistance varies with the inverse of area, not diameter, so the answer to C is incorrect. Although the answer to D is incorrect, the student has correctly carried over from the answer to C, so no further marks are lost. The answer for C should be ×4 and the answer for D should be ×8.

Correct answer

Sample	B	C	D
Length compared with A	×2	Same	×2
Diameter compared with A	Same		
Resistance compared with A	Greater ×2	Greater ×4	Greater ×4

[3 marks]

[3 marks]

Cambridge IGCSE Physics Study and Revision Guide Second Edition © Mike Folland 2016

Electrical energy and power

Electric circuits transfer energy from the battery or power source to the circuit components then into the surroundings.

You should know and be able to use the following relationships: $E = VIt$ and $P = VI$.

● Sample question

A travel kettle is designed for international use. With a 230 V supply, the power rating is 800 W.

1 Work out the current with a 230 V supply and the resistance of the element. [2 marks]

2 Find the current and power output of the kettle when used in North America with a 110 V supply. [3 marks]

3 Comment on the use of this kettle in North America. [1 mark]

Student's answer

1 $I = \dfrac{P}{V} = \dfrac{800}{230} = 3.48\ A$

$R = \dfrac{V}{I} = \dfrac{230}{3.48} = 66.1\ \Omega$ [2 marks]

2 I will stay the same.
$P = 110 \times 3.48 = 383\ W$ [1 mark]

3 The kettle will take longer to boil water. [1 mark]

Examiner's comments

1 Correct answers.

2 The current cannot stay the same because R is the same but V is different. As the student calculated the power from the wrong value of current without any further error.

3 The student made a valid comment from the calculated value of P.

Correct answers

1 $I = \dfrac{P}{V} = \dfrac{800}{230} = 3.48\ A$

$R = \dfrac{V}{I} = \dfrac{230}{3.48} = 66.1\ \Omega$ [2 marks]

2 The element is the same so R will stay the same.

$I = \dfrac{110}{66.1} = 1.66\ A$

$P = 110 \times 1.67 = 183\ W$ [3 marks]

3 The kettle will take much longer to boil water. [1 mark]

Now try this

The answer is given on p. 119.

5 A 3 kW electric heater is used to heat up 2.5 kg of water of specific heat capacity 4200 J/kg/°C. The initial water temperature is 16 °C. Work out the temperature of the water after the heater has been switched on for 2 min.

Cambridge IGCSE Physics Study and Revision Guide Second Edition © Mike Folland 2016

Electric circuits

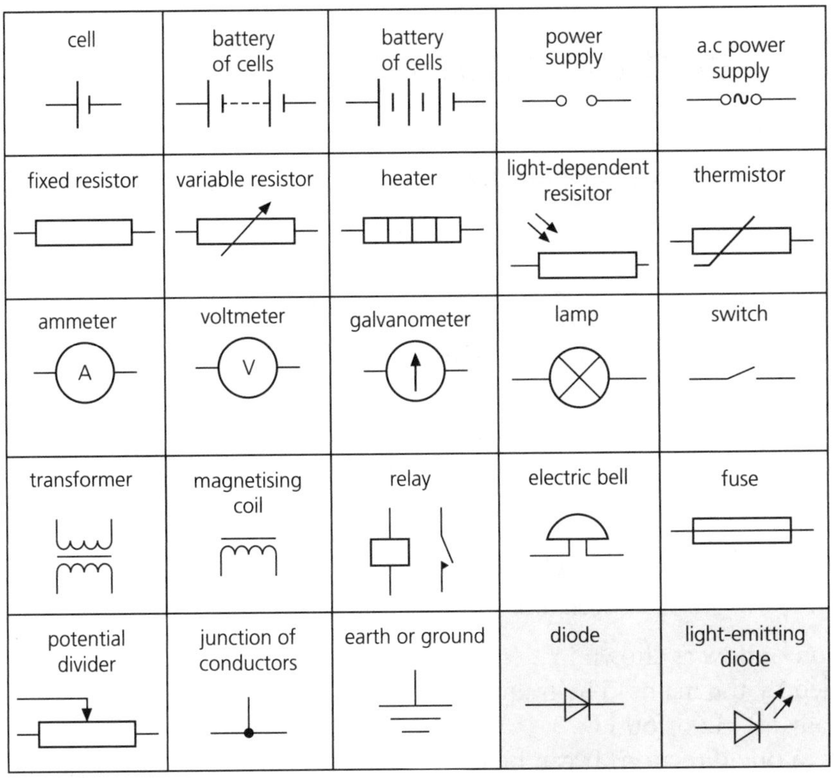

cell	battery of cells	battery of cells	power supply	a.c power supply
—⊣⊢—	—⊣┆╌╌┆⊢—	—⊣┆┆┆⊢—	—○ ○—	—○◠◡◠○—
fixed resistor	variable resistor	heater	light-dependent resisitor	thermistor
—▭—	—▱—	—▭▭—	—▭—	—▱—
ammeter	voltmeter	galvanometer	lamp	switch
—(A)—	—(V)—	—(↑)—	—⊗—	—╱—
transformer	magnetising coil	relay	electric bell	fuse
⌣⌣⌣	⌢⌢⌢	▭	⌒	—▭—
potential divider	junction of conductors	earth or ground	diode	light-emitting diode
—▭—	—•—	⏚	—▷⊢—	—▷⊢—

Figure 4.13 Circuit symbols

Examiner's tip

- You must know, be able to draw and be able to interpret all the circuit symbols shown in Figure 4.13.

● Current–voltage characteristics of some circuit components

Resistance is equal to the inverse of the gradient of the current–voltage graph of a component.

Ohmic resistor (e.g. metal wire or normal resistor at constant temperature) – resistance is constant, so gradient is constant (Figure 4.14a).

Filament lamp – resistance increases with a large increase in temperature. Temperature increases with increasing voltage and current, so gradient decreases (Figure 4.14b).

Diodes (including LEDs) – resistance is extremely high in the reverse direction, so effectively no reverse current can flow. Resistance is very low in the forward direction, so current can flow easily (Figure 4.14c).

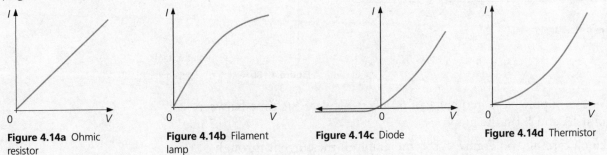

Figure 4.14a Ohmic resistor

Figure 4.14b Filament lamp

Figure 4.14c Diode

Figure 4.14d Thermistor

Thermistor (temperature-dependent resistor) – resistance decreases considerably with increasing temperature. Temperature increases with increasing voltage and current, so gradient increases.

Cambridge IGCSE Physics Study and Revision Guide Second Edition © Mike Folland 2016

Similarly the resistance of a light dependent resistor (LDR) decreases with increasing light levels.

Direct current (d.c.) and alternating current (a.c.)

Direct current always flows in one direction as shown in Figure 4.15. This is the current supplied by a battery.

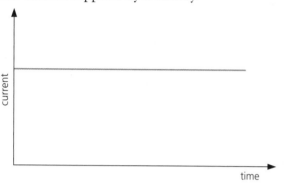

Figure 4.15 Direct current (d.c.)

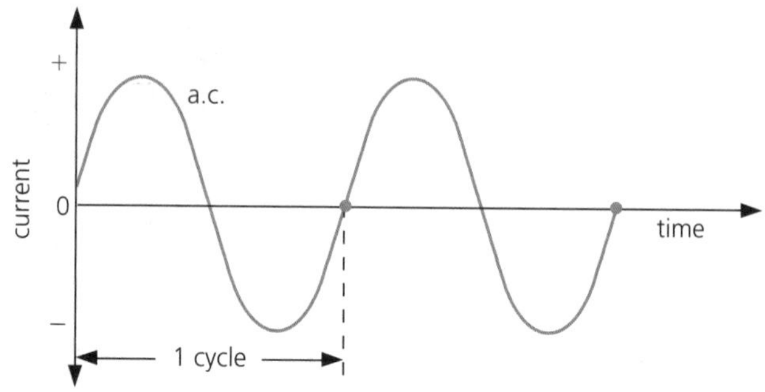

Figure 4.16 Alternating current (a.c.)

Alternating current reverses its direction of flow as shown in Figure 4.16. This is the current supplied by the mains. The frequency varies in different countries but is usually 50 Hz or 60 Hz.

A **diode** only allows current to flow in one direction. It can be used as a rectifier to convert a.c. to d.c. Figure 4.17 shows a diode used as a rectifier to recharge a battery from an a.c. supply.

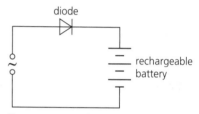

Figure 4.17 Simple charging circuit

Sample question

Figure 4.18a shows a circuit with an a.c. supply and a diode. Figure 4.18b is a graph of the voltage of terminal X compared with terminal Y.

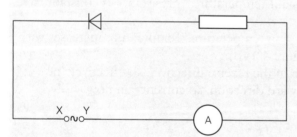

Figure 4.18a Diode circuit

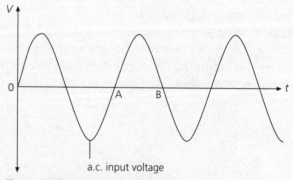

Figure 4.18b Graph of supplied voltage

1 The a.c. supply has a frequency of 60 Hz. Calculate the time between points A and B on the graph. [2 marks]

2 Sketch carefully on Figure 4.18b the graph of the current through the ammeter. [2 marks]

Student's answer

1 a.c. mains has a frequency 50 Hz.

$$time = \frac{1}{50} = 0.020 \text{ s}$$

[0 marks]

2

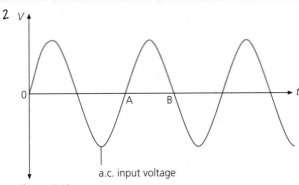

Figure 4.18c

a.c. input voltage

[1 mark]

Examiner's comments

1 The student has made two errors. First, the question was not read carefully. The student lives in a country where the mains frequency is 50 Hz. Many other countries have a mains frequency of 60 Hz and this was stated in the question. Second, the student worked out the time for a full cycle, but from A to B is a half cycle.

2 The student drew the wave pattern that would occur if the diode had been connected the other way around. However, some credit was gained for showing a half wave rectified graph.

Correct answer

1 From A to B is one half cycle.

$$time = \left(\frac{1}{2}\right) \times \left(\frac{1}{60}\right) = \frac{1}{120} = 0.0083 \text{ s (to more s.f. } t = 0.0083333 \text{ s)}$$

[2 marks]

2

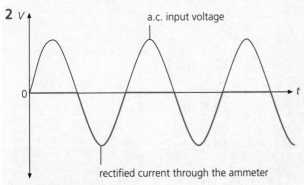

Figure 4.18d

a.c. input voltage

rectified current through the ammeter

The correct answer is shown in Figure 4.18d. Please note, no information is given about the size of the current. A graph of this shape could be drawn like this is. It would be equally acceptable with the current graph going lower or not so low. [2 marks]

Cambridge IGCSE Physics Study and Revision Guide Second Edition © Mike Folland 2016

● Series circuits

In a series circuit, there is just one path for the current to follow. The current at every point in a series circuit is the same.

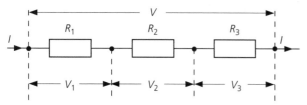

Figure 4.19

In Figure 4.19:

- I = current through R_1 = current through R_2 = current through R_3
- Total resistance = $R_1 + R_2 + R_3$

- V = p.d. across all the resistors = $V_1 + V_2 + V_3$

● Common error

✗ Current is 'used up' as it flows around a circuit.
✔ Current is a flow of electrons, which cannot be created or destroyed, so the same number must flow per second through every point in a circuit.
✔ Current is the same at every point in a series circuit.

● Sample question

In Figure 4.19, $R_1 = 4\ \Omega$ and $R_2 = 3\ \Omega$.

1 Work out the total resistance of R_1 and R_2. [2 marks]

2 The current through R_1 is 1.5 A. State the current through R_2. [2 marks]

3 Work out the voltages V_1 and V_2. [2 marks]

4 The voltage supplied across all three resistors is 12 V. Work out the value of V_3 and hence the resistance of R_3. [4 marks]

Student's answer

1 Total resistance = $4 + 3 = 7\ \Omega$ [2 marks]

2 Current through $R_2 = \frac{3}{4} \times$ current through $R_1 = 0.75 \times 1.5 = 1.125$ A [0 marks]

3 $V_1 = I_1 \times R_1 = 1.5 \times 4 = 6$ V
  $V_2 = I_2 \times R_2 = 1.125 \times 3 = 3.375$ V [2 marks]

4 $V_3 = 12 - (4 + 3) = 12 - 7 = 5$ V

$R_3 = \dfrac{V_3}{I_3} = \dfrac{5}{1.125} = 4\ \Omega$ [2 marks]

Cambridge IGCSE Physics Study and Revision Guide Second Edition © Mike Folland 2016

Examiner's comments

1 *Correct answer.*

2 *The student did not recognise that current stays the same through components in series.*

3 *The student correctly applied his answers from question 2.*

4 *The student might have been on the right lines and then made errors in substituting resistances instead of voltages. With little working and no explanation, it is impossible for the examiner to know. The answer for R_3 followed on reasonably from earlier working, so credit was gained for this.* [2 marks given]

Now try this

The answers are given on p. 119.

6 In Figure 4.19, $R_1 = 2\ \Omega$, $R_2 = 3\ \Omega$, $R_3 = 0.5\ \Omega$ and the current through the circuit = 0.8 A. Work out:
  a the supply voltage
  b the voltage across R_3.

  c R_3 is then changed for a lamp of resistance of 1.5 Ω. Work out the voltage across the lamp.

Correct answers

1 Total resistance = 4 + 3 = 7 Ω   [2 marks]

2 Current through R_2 = current through R_1 = 1.5 A

Total $R = \dfrac{1}{0.583} = 1.7\ \Omega$ [2 marks]

3 $V_1 = I \times R_1 = 1.5 \times 4 = 6$ V

$V_2 = I \times R_2 = 1.5 \times 3 = 4.5$ V   [2 marks]

4 Supply voltage = total of voltages of rest of circuit = 12 V

$12 = V_1 + V_2 + V_3$

$12 = 6 + 4.5 + V_3$

$V_3 = 12 - (6 + 4.5) = 12 - 10.5 = 1.5$ V

$R_3 = \dfrac{V_3}{I} = \dfrac{1.5}{1.5} = 1\ \Omega$ [4 marks]

● Parallel circuits

In parallel circuits, there are alternative paths for the current. The current from the source is larger than the current in each branch.

In lighting circuits in homes and businesses, lamps are connected in parallel, with the following advantages:

- Each lamp receives the full mains voltage.
- If one lamp should fail, the other lamps will continue to work.

In Figure 4.20:

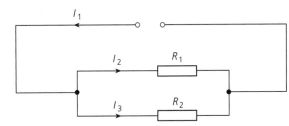

Figure 4.20 Resistors in parallel

- the total current is greater than the current in each resistor
- combined resistance of all resistors in parallel is less than resistance in any one resistor

- current from source $I_1 = I_2 + I_3$.
- combined resistance of R_1 and R_2 in parallel is given by the equation $\dfrac{1}{\text{combined resistance}} = \dfrac{1}{R_1} + \dfrac{1}{R_2}$.

Cambridge IGCSE Physics Study and Revision Guide Second Edition © Mike Folland 2016

● Sample question

In Figure 4.20, $R_1 = 4\ \Omega$, $R_2 = 3\ \Omega$, $I_1 = 4.2$ A and $I_2 = 1.8$ A.

1 Work out the current I_3. [2 marks]

2 Work out the total resistance of R_1 and R_2. [2 marks]

3 Work out the supply voltage. [2 marks]

Student's answer

1 Current $I_3 = I_1 + I_2 = 4.2 + 1.8 = 6.0$ A [0 marks]

2 Total resistance $= \dfrac{1}{R_1} + \dfrac{1}{R_1} = \dfrac{1}{4} + \dfrac{1}{3} = 0.25 + 0.333 = 0.583\ \Omega$ [0 marks]

3 Supply voltage $= 6.0 \times 0.583 = 3.50$ V [2 marks]

Examiner's comments

1 The student has wrongly thought that I_3 is the total current.

2 The student has applied the wrong equation.

3 Full marks are given despite the wrong answer. The student correctly followed on from questions 1 and 2.

Correct answers

1 Current $I_3 = I_1 - I_2 = 4.2 - 1.8 = 2.4$ A [2 marks]

2 $\dfrac{1}{\text{Total } R} = \dfrac{1}{R_1} + \dfrac{1}{R_2} = \dfrac{1}{4} + \dfrac{1}{3} = 0.25 + 0.333 = 0.583\ \Omega$ [2 marks]

3 Supply voltage $= 2.4 \times 3 = 7.2$ V [2 marks]

Now try this

The answers are given on p. 119.

7 In Figure 4.20, $R_1 = 2\ \Omega$, $R_2 = 3\ \Omega$ and the supply voltage = 6 V. Work out:
  a I_2
  b I_3
  c I_1
  d the total resistance.

A **potential divider**, or potentiometer, provides a voltage that varies with the values of two resistors in a circuit.

Figure 4.21 shows a potential divider with two separate resistors. If the value of R_1 or R_2 changes, the output voltage will change.
  If R_1 increases with R_2 unchanged, the output voltage decreases.
  If R_2 increases with R_1 unchanged, the output voltage increases.

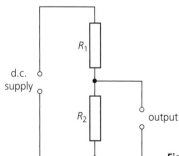

Figure 4.21 Potential divider with two separate resistors

Now try this

The answers are given on p. 119.

8 In Figure 4.21, the supply voltage is 6 V and R_1 is replaced by a thermistor. The value of R_2 is much less than the greatest resistance of the thermistor and much more than the lowest resistance of the thermistor. What is the output voltage when the thermistor is
  a very hot
  b very cold?

Cambridge IGCSE Physics Study and Revision Guide Second Edition © Mike Folland 2016

A potential divider can also be made as a single component, sometimes called a rheostat. When the slider moves up and down the resistor in Figure 4.22, the resistance above and below the slider changes, so the output voltage changes. As the slider moves down, the output voltage decreases to zero. As the slider moves up, the output voltage increases to the value of the supply voltage.

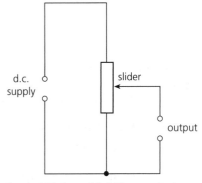

Figure 4.22 Potential divider as a single component

The resistance of a **light-dependent resistor (LDR)** falls with increasing light level. It can be connected in a circuit that is required to respond to changes in light level.

The resistance of a thermistor decreases considerably with increasing temperature.

Figure 4.23 shows a circuit that acts as a fire alarm. When the temperature of the thermistor rises, its resistance falls. The thermistor and fixed resistor R are a potential divider, so the p.d. between S and T rises and enough current flows into the relay for it to switch on the bell.

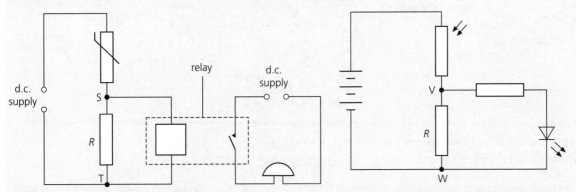

Figure 4.23 Fire alarm circuit

Figure 4.24 Light-sensitive circuit using a remote light-emitting diode

Figure 4.24 shows a circuit that acts as a warning when too much light enters an automated photographic laboratory. A light-emitting diode (LED) on the control panel outside the laboratory can light up to show the warning.

When operating correctly in the dark, the resistance of the LDR is high. The p.d. between V and W is low so no current flows through the LED.

If light enters the laboratory, there is an increase in light level and the resistance of the LDR falls. The LDR and fixed resistor R are a potential divider, so the p.d. between V and W rises and enough current flows through the LED for it to light up and give a warning.

Cambridge IGCSE Physics Study and Revision Guide Second Edition © Mike Folland 2016

Digital electronics

In an **analogue** circuit, the voltages can vary continuously to take any value in the available range.

In a **digital** circuit, voltage can be in only one of two states:

1 zero voltage denoted by OFF, LOW or binary 0

2 maximum available voltage denoted by ON, HIGH or binary 1.

Logic gates are digital components with one or more inputs, which determine the value of the one output. Logic gates are complex devices that contain many components.

The truth table for a logic gate shows the relationship between the input and output binary values.

Examiner's tip

You must be able to describe the action of the NOT, OR, AND, NOR and NAND gates shown on the left with their truth tables, as well as know and be able to use their symbols.

A NOT gate reverses the value of its single input.

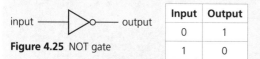

Figure 4.25 NOT gate

Input	Output
0	1
1	0

An OR gate has a HIGH output if either or both of the inputs are HIGH.

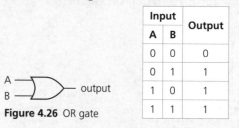

Figure 4.26 OR gate

Input		Output
A	B	
0	0	0
0	1	1
1	0	1
1	1	1

An AND gate has a HIGH output only if both the inputs are HIGH.

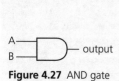

Figure 4.27 AND gate

Input		Output
A	B	
0	0	0
0	1	0
1	0	0
1	1	1

A NOR gate is the same logically as an OR gate followed by a NOT gate.

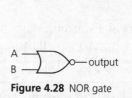

Figure 4.28 NOR gate

Input		Output
A	B	
0	0	1
0	1	0
1	0	0
1	1	0

A NAND gate is the same logically as an AND gate followed by a NOT gate.

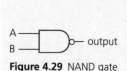

Figure 4.29 NAND gate

Input		Output
A	B	
0	0	1
0	1	1
1	0	1
1	1	0

Cambridge IGCSE Physics Study and Revision Guide Second Edition © Mike Folland 2016

You must be able to design simple digital circuits. The circuit in Fig 4.30 is to meet this situation. A student requires an OR gate but has a stock of only NOR gates and a NOT gate. She designs this circuit.

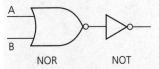

A

B

NOR   NOT

Figure 4.30 Combining NOR and NOT gates to make an OR gate

Truth table for the circuit in Figure 4.30:

Input to NOR		Output from NOR	Output from NOT
A	**B**		
0	0	1	0
0	1	0	1
1	0	0	1
1	1	0	1

The whole circuit acts as an OR gate.

● Sample question

1 Draw the truth table for the circuit shown in Figure 4.31. [3 marks]

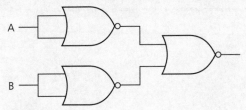

A

B

Figure 4.31

2 Which circuit element is equivalent to the circuit shown in Figure 4.31? [1 mark]

Student's answer

1
Input		Output
A	**B**	
0	0	0
0	1	1
1	0	1
1	1	1

[0 marks]

2 *The circuit acts as an OR gate.* [1 mark]

Examiner's comments

1 The student failed to label intermediate points in the circuit and include them in the truth table. The second and third lines in the table contain errors and without working the student can be given little credit.

2 Although the answer is wrong, the deduction from the student's truth table was correct.

Cambridge IGCSE Physics Study and Revision Guide Second Edition © Mike Folland 2016

Correct answer

1 Label the inputs to the right-hand NOR gate as C (upper) and D (lower).

Input		C	D	Output
A	B			
0	0	1	1	0
0	1	1	0	0
1	0	0	1	0
1	1	0	0	1

[3 marks]

2 The circuit acts as an AND gate. [1 mark]

Now try this

The answers are given on p. 119.

9 Copy and complete the truth table for the circuit shown in Figure 4.32.

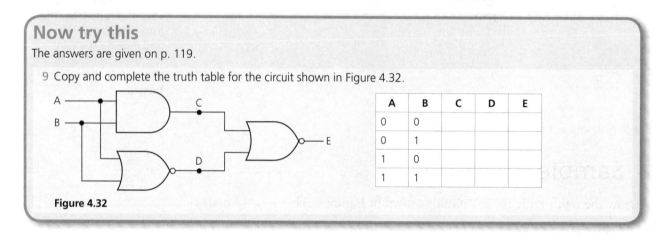

A	B	C	D	E
0	0			
0	1			
1	0			
1	1			

Figure 4.32

● Dangers of electricity

Some common hazards are outlined below:

- Damaged insulation can lead to very high currents flowing in inappropriate places. This poses a danger of electric shock or fire.
- Cables that overheat owing to excessive current can lead to fire or damage in the appliance or in the cables and insulation.
- Water lowers the resistance to earth, so damp conditions can lead to the current shorting and can cause shocks. Electrical devices for use in damp conditions must be designed to high standards of damp proofing, especially connectors and switches.

A fuse is a piece of wire that melts ('blows') when too much current flows through it. This switches off the circuit to protect against shock, fire or further damage.

A fuse with the appropriate melting current must be chosen for each application.

A circuit breaker is a type of relay that is normally closed, but opens ('trips') when too much current flows through it. Circuit breakers react much faster than fuses and have increasingly replaced them in consumer units. The appropriate current setting must be chosen for the application.

Devices with metal cases should have the case connected to earth (ground). If there is a fault which causes the case to become live, a high current can safely flow to earth and cause the fuse to 'blow' or the circuit breaker to 'trip'.

● Sample question

People are gathered after dark on wet grass. Explain whether the following three situations are potentially dangerous:

1 A heater and several high-powered electric lamps are supplied by an old extension cable. [2 marks]

2 There is a cut in the outer insulation of the cable. [2 marks]

3 The devices are connected to a switch lying on the lawn. [2 marks]

Student's answers

1 The electrical power is likely to require too much current in the cable, leading to overheating. This could cause a fire or melting of the insulation. [2 marks]

2 There is insulation on the individual wires, so the cable is safe. [0 marks]

3 The dew on the cable connection could cause an electric shock. [2 marks]

Examiner's comments

1 Correct answer.

2 Incorrect answer – using a cable with any sort of cut is unsafe practice.

3 Correct answer.

Correct answers

1 The electrical power is likely to require too much current in the cable, leading to overheating. This could cause a fire or melting of the insulation. [2 marks]

2 There could also be a cut in the insulation of the individual wires, which would be difficult to see. There would be a danger of electric shock. [2 marks]

3 The water on the cable switch could cause an electric shock. [2 marks]

Cambridge IGCSE Physics Study and Revision Guide Second Edition © Mike Folland 2016

Electromagnetic induction

When the magnetic field through a circuit changes, an e.m.f. (voltage) is induced.

In Figure 4.33, voltage is induced only when the wire moves upwards (direction 1) or downwards (direction 2). The meter deflects in opposite directions in these two cases, but only when the wire is in motion. When moved in the other directions, the wire does not cut the magnetic field, so no voltage is induced.

Figure 4.34 is an example of an electromagnet where the magnet moves but the wires are stationary.

● Common error

✘ A **current** is induced when the magnetic field through a circuit changes.

✔ A **voltage or e.m.f.** is induced when the magnetic field through a circuit changes.

✔ A current often flows as a result of the induced voltage.

The induced e.m.f. increases with an increase in:

● speed of relative motion of the magnet and circuit
● number of turns of any coil or circuit
● strength of the magnet.

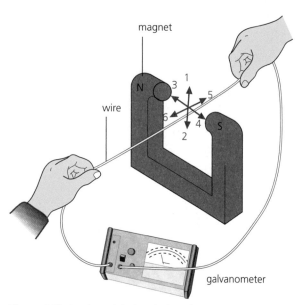

Figure 4.33 A voltage is induced when the wire is moved up or down in the magnetic field

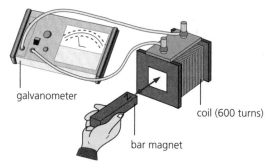

Figure 4.34 A voltage is induced in the coil when the magnet is moved in or out

The direction of the induced e.m.f. *opposes* the change that caused it. In Figure 4.35, the moving magnet induces an e.m.f. in the coil, which causes a current to flow through the coil. The current produces a magnetic field, which opposes the movement of the magnet.

When the magnet moves down in Figure 4.35a, the top of the coil becomes an N pole. When the magnet moves up in Figure 4.35b, the top of the coil becomes an S pole.

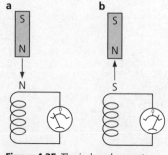

Figure 4.35 The induced current opposes the motion of the magnet

It is illegal to photocopy this page

Cambridge IGCSE Physics Study and Revision Guide Second Edition © Mike Folland 2016

Now try this

The answers are given on p. 119.

10 The magnet in Figure 4.36 is released and falls away from the coil.

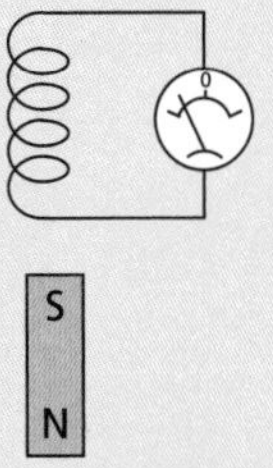

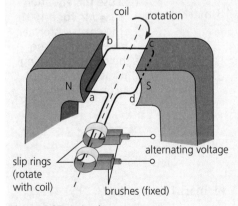

Figure 4.36

 a Explain how the falling magnet causes the coil to become magnetised.
 b State and explain the polarity of the coil when magnetised.

You should be able to state and use the relative directions of motion, field and induced e.m.f.

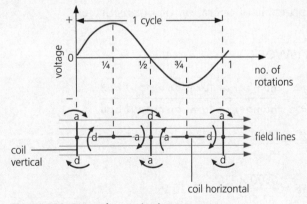

Figure 4.37 The right-hand generator rule

● Simple a.c. generator

Figure 4.38 shows a simple a.c. generator with a coil rotating within the field of a magnet. An e.m.f. is induced in the coil connected to the output terminals through slip rings on the axle, which rotate with the coil, and fixed carbon brushes. The wires on each side of the coil cut the field alternately moving up and down, so the e.m.f. is induced in alternating directions, as shown by the graph in Figure 4.39.

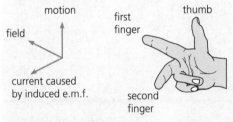

Figure 4.38 A simple a.c. generator

Figure 4.39 Output from a simple a.c. generator

Examiner's tip

You must be able to relate the peaks and zeroes of the voltage output in the upper diagram of Figure 4.39 to the position of the coil shown in the lower diagram.

Cambridge IGCSE Physics Study and Revision Guide Second Edition © Mike Folland 2016

Transformers

A transformer transforms (changes) an alternating voltage from one value to another of greater or smaller value of the same frequency.

Figure 4.40 shows a transformer with two coils wound on the same soft iron core. The primary coil is supplied with an a.c. and the secondary coil provides a.c. to another circuit.

The primary coil acts as an electromagnet and produces an alternating magnetic field in the soft iron core. The secondary coil is in this alternating magnetic field, so an alternating e.m.f. is induced.

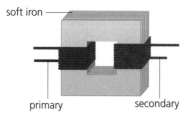

soft iron

primary   secondary

Figure 4.40 Primary and secondary coils of a transformer

● Common error

✖ Transformers work with a.c. and d.c.
✔ Transformers work only with a.c.

A step-up transformer has more turns on the secondary than the primary coil and V_s is greater than V_p. In a step-down transformer, there are fewer turns on the secondary than the primary coil and V_s is less than V_p.

● Sample question

A transformer used by students in a school laboratory has 5500 turns on the input coil and is supplied with 110 V a.c. The output coil has 500 turns.

1 Calculate the output voltage.   [3 marks]

2 Describe the principle of operation of a transformer.   [3 marks]

Student's answer

1 $\dfrac{output\ voltage}{input\ voltage} = \dfrac{number\ of\ turns\ on\ output\ coil}{number\ of\ turns\ on\ output\ coil}$

$\dfrac{output\ voltage}{110} = \dfrac{5500}{500}$

$output\ voltage = \dfrac{5500 \times 110}{500} = 1210\ V$ [1 mark]

2 The primary coil acts as an electromagnet supplied with an alternating current. This flows in the soft iron core. Therefore, the secondary coil has an alternating current.   [1 mark]

Examiner's tips

• Know and be able to use the equation $\dfrac{V_p}{V_s} = \dfrac{N_p}{N_s}$ where V_p and V_s are the voltages across the primary and secondary coils and N_p and N_s are the numbers of turns of the primary and secondary coils.

• Know and be able to use the equation $I_p V_p = I_s V_s$ for a 100% efficient transformer, where I_p and I_s are the currents flowing in the primary and secondary coils.

Cambridge IGCSE Physics Study and Revision Guide Second Edition © Mike Folland 2016

Examiner's comments

1 *The student started with the correct equation but muddled up the numbers of turns on the two coils. In addition, the student should have realised that students in a laboratory would never have access to such a high voltage, so something must have gone wrong in the calculation.*

2 *The student knows that there is a magnetic field but completely fails to use this information. The answer also gives the impression that current flows through the core to the secondary coil, which is completely wrong.*

Correct answer

1 $\dfrac{\text{output voltage}}{\text{input voltage}} = \dfrac{\text{number of turns on output coil}}{\text{number of turns on input coil}}$

$\dfrac{\text{output voltage}}{110} = \dfrac{500}{5500}$

output voltage $= \dfrac{500 \times 110}{5500} = 10$ V   [3 marks]

2 The primary coil acts as an electromagnet supplied with an alternating current. This produces an alternating magnetic field in the soft iron core. The secondary coil is in this alternating magnetic field, so an alternating e.m.f. is induced.   [3 marks]

Electricity is transmitted over large distances at very high voltages in order to reduce the energy losses due to the resistance of the transmission lines. This is achieved by having a step-up transformer at the power station to increase the voltage to several hundred thousand volts. Where the electricity is to be used, there is a series of step-down transformers to reduce the voltage to values suitable for use in factories or homes.

> Power loss in a cable = p.d. across length of cable × current (I)
> For a given cable of fixed resistance:
> p.d. = I × resistance of cable (R)
> Power loss = $I \times R \times I = I^2R$
> Power is therefore transmitted at the highest possible voltage in order to reduce the current and thus the losses in the cables.

Now try this

The answers are given on p. 119.

11 A transformer is used to provide an a.c. 6 V supply for a laboratory from 240 V a.c. mains. The secondary coil of the transformer has 100 turns. Work out how many turns the primary coil should have.

Electromagnetism

A wire or coil carrying an electric current produces a magnetic field. The higher the current, the stronger the magnetic field. If the current reverses, the direction of the magnetic field is also reversed.

For the field due to a straight wire, the greater the distance from the wire, the weaker the magnetic field.

A solenoid is a long cylindrical coil. When a current flows, the field pattern outside the solenoid is similar to that of a bar magnet. Inside the solenoid there is a strong field parallel to the axis. The right-hand grip rule gives the direction of the field.

The fingers of the right hand grip the solenoid pointing in the direction of the current and the thumb points to the N pole.

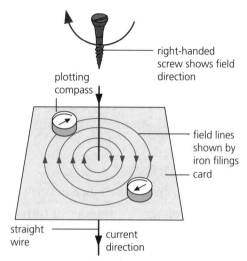

Figure 4.41 Field due to a straight wire

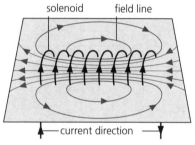

Figure 4.42a Field due to a solenoid

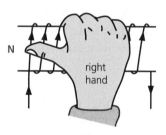

Figure 4.42b The right-hand grip rule

The closeness of the field lines in Figure 4.42a indicate the strength of the field. The field is very strong within the solenoid. Outside the solenoid, the further away a point is, the weaker the field.

The strength of the magnetic field of a solenoid is greatly increased if there is a core of soft iron inside the coils. As the field can be switched on and off with the current, this is the basis for the **electromagnet** (Figure 4.43).

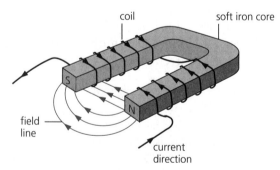

Figure 4.43 Horseshoe electromagnet

Cambridge IGCSE Physics Study and Revision Guide Second Edition © Mike Folland 2016

Uses of electromagnets

Electromagnets are used in cranes to lift iron objects and scrap iron, as well as in many electrical devices. Figure 4.44 shows an electric bell.

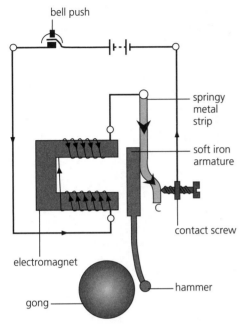

Figure 4.44 Electric bell

When the bell push is pressed, a current flows through the electromagnet. The armature is attracted to the electromagnet and the hammer strikes the gong. The movement of the armature breaks the circuit that applies current to the electromagnet. The armature is released and springs back. The circuit is re-made, the process repeats and the bell rings continually for as long as the bell push is pressed.

Sample question

For the electric bell shown in Figure 4.44:

1 explain why soft iron is used for the armature

2 choose a suitable material for the core of the electromagnet. Give your reasons. [3 marks]

Student's answer

1 *A strong magnet is required.* [0 marks]
2 *Use soft iron, as the electromagnet must be switched on and off repeatedly.* [2 marks]

Examiner's comments

1 The armature is not a permanent magnet but becomes an induced magnet only when the electromagnet produces a field. When no current flows, the armature must spring back, so there must be no permanent attraction to the core of the electromagnet. Therefore, soft iron should be used.

2 Correct answer.

Examiner's tips

- You need to be able to describe applications of the magnetic effects of a current, including the action of a relay (described below).

- You should also be able to show understanding of the use of a relay in switching circuits.

- You should also be able to describe the operation of other applications, e.g. electromagnet, electric bell and moving-coil loudspeaker.

Correct answer

1 Soft iron is used because the armature must be attracted to the electromagnet only when the current is switched on. [1 mark]

2 Use soft iron, as the electromagnet must be switched on and off repeatedly. [2 marks]

● The magnetic relay

The relay is a device that enables one electric circuit to control another. It is often used when the first circuit carries only a small current (e.g. in an electronic circuit) and the second circuit requires a much higher current.

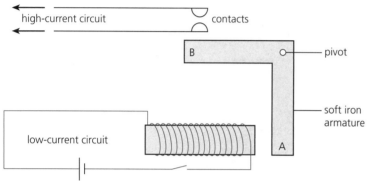

Figure 4.45 Magnetic relay

When the switch is closed in the low-current circuit in Figure 4.45, current flows to the electromagnet, which attracts end A of the soft iron armature. The armature pivots and end B moves up to close the contacts in the high-current circuit. This circuit is now complete and the high current flows through the device, e.g. a motor, a heater or an alarm bell.

Now try this

The answers are given on p. 119.

12 A loudspeaker is made up essentially of a stationary magnet that is close to a small coil fixed to a paper cone. The signal from the amplifier is a small alternating current supplied to the coil. Describe briefly:
   a the variation of the magnetic field produced by the coil
   b the variation of the magnetic force on the coil
   c the motion of the paper cone.

Cambridge IGCSE Physics Study and Revision Guide Second Edition © Mike Folland 2016

The motor effect

A wire or conductor carrying a current in a magnetic field experiences a force. This can be demonstrated with the apparatus shown in Figure 4.46.

The loosely suspended wire will be seen to move up when the current is switched on. The wire will move down if either the current is reversed or the magnet poles are swapped to reverse the field. If both the field and current are reversed, the wire will again move up.

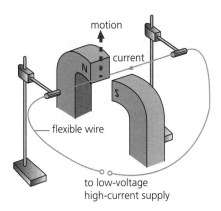

Figure 4.46 Demonstrating the motor effect

● The motor rule

You must be able to state and use the left-hand motor rule to determine the relative directions of force, field and current.

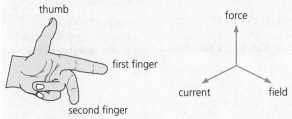

Figure 4.47 The left-hand motor rule

● Sample question

Figure 4.48 shows a wire in a magnetic field. The current through the wire is switched on.

1 State and explain the direction of the force on the wire when the current is switched on.

2 For each of the following changes, made one at a time, state whether the magnitude of the force on the wire increases, stays the same, decreases or decreases to zero:

   a current changes direction

   b current drops to zero

   c current increases

   d magnetic field increases

   e magnetic field changes direction.

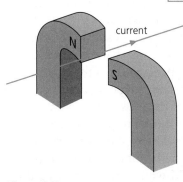

Figure 4.48

Cambridge IGCSE Physics Study and Revision Guide Second Edition © Mike Folland 2016

Student's answer

1 *Force is up, as in Figure 4.46.*

2 a *Changes direction.*
  b *Becomes zero.*
  c *Increases.*

d *Increases.*
e *Changes direction.*

Examiner's comments

1 The student did not observe that the current is reversed from Figure 4.46.

2 The student seems in places to have committed the 'classic error' of answering the question that was expected not what was asked.

a The student answered a different question.

b Correct answer.

c Correct answer.

d Correct answer.

e The student answered a different question.

Correct answer

1 Force is down by left-hand motor rule.

2 a Force stays the same.

  b Force decreases to zero.

  c Force increases.

d Force increases.

e Force stays the same.

Examiner's tip
Always answer the question actually asked, not the question you expect.

● Deflection of beams of charged particles

A beam of charged particles is equivalent to a current and experiences a similar force in a magnetic field. Figure 4.49 shows an experiment with a beam of positively charged particles in a Maltese cross tube. Those particles that miss the cross cause the fluorescent screen to glow, showing a shadow pattern in the shape of the cross. Two bar magnets are placed each side of the tube to produce a horizontal magnetic field.

The shadow pattern moves down as the particles in the beam experience a force from the magnetic field. Moving positively charged particles are equivalent to a current in their direction of travel and the left-hand motor rule is used to find the direction of deflection.

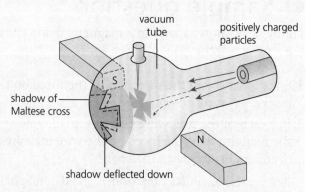

Figure 4.49 Deflection of a beam of charged particles

- First finger points in direction of field from N pole to S pole.
- Second finger points from right to left parallel with motion of the positively charged particles.
- Thumb points down to show direction of deflection.

● Common error

✖ A beam of moving electrons is equivalent to a current flowing in the same direction as the electrons.
✔ Electrons are negatively charged, so the current is in the opposite direction to the electron flow.
✔ A beam of positively charged particles is equivalent to a current flowing in the same direction as the motion of the particles.

● Sample question

Figure 4.50 shows the fluorescent screen of an electron tube and the shadow of a Maltese cross when electrons travel towards the viewer in the absence of any magnetic field.

Figure 4.50

State and explain how this shadow will move when two bar magnets are placed in the positions shown. [4 marks]

Student's answer

The magnetic field is from left to right. [1 mark] *The electrons move out of the paper.* [0 marks] *Using the left-hand rule, the beam and shadow move upwards.* [2 marks]

Examiner's comments

The student's statement about the magnetic field is correct, but electrons are negatively charged, so when they move in one direction, the corresponding current is in the opposite direction. The student correctly applied the left-hand rule to their directions, although the current direction was wrong.

Correct answer

The magnetic field is from left to right. [1 mark] The electrons move towards the observer, so the current moves into the paper. [1 mark] Using the left-hand rule, the beam and shadow move downwards. [2 marks]

Cambridge IGCSE Physics Study and Revision Guide Second Edition © Mike Folland 2016

The d.c. motor

Figure 4.51 shows the coil of a simple d.c. motor. The coil in a magnetic field carries a current. The directions of the forces on the coil are worked out using the left-hand motor rule. There is a force from the magnetic field on the wire **ab** in one direction (upwards). The current in the wire **cd** is in the other direction, so it experiences a force in the opposite direction (downwards). These two forces cause the coil of wire to turn. The turning effect is increased by:

- increasing the number of turns on the coil
- increasing the current
- increasing the strength of the magnetic field.

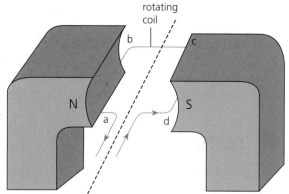

Figure 4.51 The principle of a simple d.c. motor

This turning effect causes the coil of a d.c. motor to rotate continuously. The commutator and brushes act as a switching mechanism that changes the direction of the current every half turn to allow continuous rotation. When the coil has turned through half a turn, wire **ab** is now on the right, so moves down. Similarly, wire **cd** is now on the left and moves up.

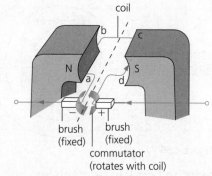

Figure 4.52 The commutator and brushes of a d.c. motor

Now try this

The answers are given on p. 119.

13 a Figure 4.53 shows a coil with several turns carrying a current in the magnetic field shown. State the effect that the current has on the coil, including any direction.
   b State whether the size of this effect is increased, the same, decreased or decreased to zero when:
   (i) the current is reversed
   (ii) the current is increased
   (iii) the magnets are removed
   (iv) the number of turns is increased.

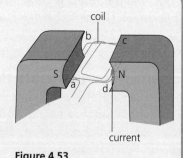

Figure 4.53

14 Figure 4.54 shows a coil that can rotate in a magnetic field.
   a State the direction of any forces on:
   (i) wire ab
   (ii) wire cd.
   b The coil is rotated so that it is vertical with wire ab uppermost. State the direction of any forces now acting on:
   (i) wire ab
   (ii) wire cd.
   Explain how you reached your answers.
   c The coil is rotated so that it is again horizontal with wire ab on the right and wire cd on the left. State the direction of any forces now acting on:
   (i) wire ab
   (ii) wire cd.

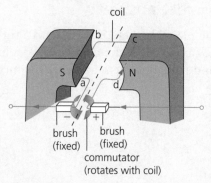

Figure 4.54

Cambridge IGCSE Physics Study and Revision Guide Second Edition © Mike Folland 2016

Atomic physics

> ## Main points to revise
>
> - how to describe the structure of an atom and its nucleus
> - how to use the term 'isotope'
> - how to use nuclide notation
> - the basic characteristics of the three types of radioactive emissions
> - how to perform half-life calculations
> - how to perform half-life calculations taking into account background radiation
> - how to describe precautions for handling, using and storing radioactive materials safely
> - how to write radioactive decay equations using nuclide notation.

● Key terms

Term	Definition
Atom	The smallest possible unit of an element
Electron	A very light negatively charged particle found in an atom
Proton	A positively charged subatomic particle found in the nucleus of an atom
Neutron	An uncharged subatomic particle found in the nucleus of an atom
Isotope	One form of an element that has a different number of neutrons in the nucleus from other isotopes of the same element
α-particle	A helium nucleus made up of two protons and two neutrons
β-particle	A high-speed electron emitted by a nucleus
γ-ray	A high-frequency electromagnetic wave
Half-life	The average time for half the atoms in a radioactive sample to decay

Atomic structure

The atom is the smallest particle of an element. It is made up of a central nucleus, with all the positive charge and nearly all of the mass, and negatively charged electrons in orbit. The nucleus is very much smaller than the electron orbits, so the majority of every atom is empty space.

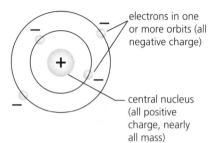

Figure 5.1 The nuclear atom

The nuclear atom was confirmed by observing a beam of α-particles (positively charged particles) travelling towards a thin metal foil. The vast majority passed straight through without being deflected. However, a few α-particles were deflected, some through a large angle, and a very small proportion bounced back (see Figure 5.2).

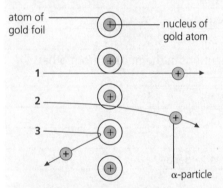

Figure 5.2 Scattering of α-particles by thin gold foil

Path 1 is a long way from any nucleus and the α-particle is undeflected.
Path 2 is close to a nucleus and there is some deflection.
Path 3 heads almost straight for a nucleus and the α-particle rebounds back.

● Nucleus

A nucleus contains protons and neutrons, together called **nucleons**.

A proton is about 2000 times more massive than an electron and has a positive charge of the same magnitude as an electron's negative charge.

A neutron has about the same mass as a proton but has no charge.

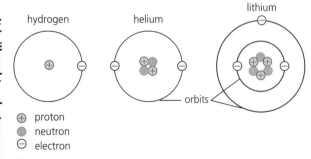

Figure 5.3 Protons, neutrons and electrons in atoms

Cambridge IGCSE Physics Study and Revision Guide Second Edition © Mike Folland 2016

The number of protons in a nucleus is called the **proton number** (Z) and is the same as the number of electrons in orbit.

The number of neutrons is about the same as the number of protons for light elements and rises to about one and a half times the number of protons for the heaviest elements.

The number of nucleons (protons and neutrons) is called the **nucleon number** (A).

The nuclide (type of nucleus) of an element can be written with the notation $_Z^A X$, where X is the chemical symbol for the element.

● Isotopes

Isotopes of the same element are different forms that have the same number of protons but different numbers of neutrons in the nucleus.

● Sample question

The normally occurring isotope of carbon is carbon-12, written $_6^{12}C$ in nuclide notation.

1 Write down the nucleon and proton numbers of carbon-12.   [2 marks]

2 Write down the number of electrons in a neutral atom of carbon-12.   [1 mark]

3 Carbon-14 is a radioactive isotope that exists in small quantities in the atmosphere. Write down the nucleon and proton numbers of carbon-14.   [1 mark]

4 Write down the nuclide notation for carbon-14.   [2 marks]

5 Work out the number of neutrons in a nucleus of carbon-14 and state the difference between the nuclei of carbon-12 and carbon-14.   [2 marks]

Student's answer

1 Nucleon number, $A = 12$
   Proton number, $Z = 6$ [2 marks]
2 Number of electrons = 12   [0 marks]
3 Nucleon number, $A = 14$
   Proton number, $Z = 6$ [1 mark]
4 $_{14}^{6}C$ [0 marks]
5 A carbon-14 nucleus is bigger, with more particles.   [0 marks]

Examiner's comments

1 Correct answers.

2 The student has incorrectly thought that the number of electrons is the same as the number of nucleons.

3 Correct answers.

4 The student has mixed up the nucleon and proton numbers – care needs to be taken here!

5 The first part of the question has not been answered and the rest of the answer is too vague.

Cambridge IGCSE Physics Study and Revision Guide Second Edition © Mike Folland 2016

Correct answers

1 Nucleon number, A = 12

Proton number, Z = 6   [2 marks]

2 Number of electrons = number of protons = 6   [1 mark]

3 Nucleon number, A = 14

Proton number, Z = 6   [1 mark]

4 $^{14}_{6}C$

5 Number of neutrons in carbon-14 nucleus = nucleon number – proton number = $A - Z$ = 14 – 6 = 8

Number of neutrons in carbon-12 nucleus = 12 – 6 = 6

A carbon-14 nucleus has two extra neutrons.   [2 marks]

● **Common errors**

✖ Mixing up the positions of the nucleon and proton numbers in nuclide notation, e.g. $^{2}_{4}He$

✔ $^{4}_{2}He$ has nucleon number A = 4, and proton number Z = 2.

✖ Writing one or both of the numbers to the right of the element symbol, as in some old books, e.g. $^{4}He_{2}$

✔ $^{4}_{2}He$ has both numbers to the left of He.

Some radioactive isotopes occur naturally, e.g. carbon-14 is produced in the atmosphere by cosmic rays.

Many radioactive isotopes are produced artificially in nuclear reactors and have a wide range of practical uses, e.g. as a source of radiation to kill cancers (see p. 115) or as tracers in the human body or in a pipeline.

Now try this

The answers are given on p. 119.

1 Copy and complete the table to indicate the composition of an atom of each of the isotopes of strontium given.

Isotope	Number of protons	Number of neutrons	Number of electrons
$^{88}_{38}Sr$			
$^{90}_{38}Sr$			

Examiner's tip

- You do not have to remember the equations in nuclide notation in this topic. They are given as examples.

- Exam questions regularly include nuclide equations (see the sample question on Radioactivity, p. 113). You do need to be able to complete and balance these equations.

Nuclear fission and fusion

Fission occurs when a nucleus of certain 'heavy' isotopes is bombarded by a neutron of the appropriate velocity and splits into two 'lighter' nuclei, releasing two neutrons and a large amount of energy.

e.g. $^{235}_{92}U + ^{1}_{0}n \rightarrow ^{144}_{56}Ba + ^{90}_{36}Kr + 2^{1}_{0}n$

This reaction occurs in a nuclear reactor or atomic bomb. The neutrons released bombard and cause further uranium atoms to split in a chain reaction. In an atomic bomb, this is uncontrolled and leads to an explosion. In a nuclear reactor, the number of neutrons is carefully controlled. The 'lighter' nuclei produced are themselves highly radioactive nuclear waste, which is difficult and expensive to dispose of.

Fusion occurs under conditions of extremely high temperature and pressure when light nuclei can join together. This forms a nucleus of an element with a higher proton number and releases a large amount of energy.

e.g. $^{1}_{1}H + ^{2}_{1}H \rightarrow ^{3}_{2}He$

This is the reaction that occurs in the Sun and other stars, as well as in the hydrogen bomb. Research reactors are currently experimenting into ways of maintaining controlled fusion reactions for possible power stations of the future

Now try this

The answers are given on p. 120.

2 a Explain the difference between nuclear fission and nuclear fusion.
  b State one significant similarity between them.
  c State where a fission reaction occurs.
  d State where a fusion reaction occurs.

Radioactivity

Radioactivity occurs when an unstable nucleus decays and emits one or more of the three types of radiation: α-(alpha) particles, β-(beta) particles or γ-(gamma) rays.

In collisions between radioactive particles and molecules in the air, the radioactive particles knock electrons out of the atoms, leaving the molecules positively charged. This is called **ionisation**.

Radioactivity is a random process. It is impossible to know when an individual radioactive nucleus will decay. A large sample will, on average, decay in a regular pattern, but there will always be short-term variations in the rate of decay.

Background radiation occurs naturally, mainly due to radioactivity in rocks and the air and particles from space called cosmic rays. Like all radioactive processes, the intensity of background radiation fluctuates in the short term.

Radioactivity can be detected by a 'Geiger counter' or more precisely a Geiger–Müller tube connected to a counter. Radiation causes a pulse of current to flow between the electrodes, which flows to the counter or ratemeter and often to a loudspeaker to produce the characteristic 'clicking'.

● α-particles, β-particles and γ-rays

Emission	Nature	Charge	Penetration	Ionising effect
α-particle	Helium nucleus (two protons and two neutrons)	+2	Stopped by thick paper or a few centimetres of air	Very strong
β-particle	High-speed electron	−1	Stopped by a few millimetres of aluminium	Weak
γ-ray	Electromagnetic radiation	None	Only stopped by many centimetres of lead	Very weak

Now try this

The answers are given on p. 120.

3 The penetration of two radioactive samples is tested by measuring the count rate with various types of shielding between the sample and the counter. The numbers in the table below indicate the count rate (CR) with each type of shielding in place – no shielding, thick card, 3 mm of aluminium (Al), 20 cm of lead (Pb).

Copy the table and tick the appropriate boxes in the right-hand three columns to show the type or types of emission from that sample.

Sample	CR (none)	CR (card)	CR (Al)	CR (Pb)	α	β	γ
1	6000	1000	1000	20			
2	3000	3000	20	20			

Cambridge IGCSE Physics Study and Revision Guide Second Edition © Mike Folland 2016

Smoke alarm

Americium-241 is an α-emitter with a long half-life, which is used in smoke alarms. The nuclide equation for its decay is:

$$^{241}_{95}\text{Am} \rightarrow\ ^{237}_{93}\text{Np} +\ ^{4}_{2}\text{He}$$ (Please note, a He nuclide is the same as an α-particle.)

In the absence of smoke, the α-particles ionise the air and a small current flows between electrodes. In the presence of smoke, the ions cannot flow freely between the electrodes. The reduction of current is detected by a circuit, which sounds the alarm.

Thickness gauges

A β- or γ-emitter is placed on one side of a strip of paper, plastic or metal during its manufacture. A radiation detector on the other side of the strip monitors how much has been absorbed, which indicates the thickness of the strip. β-Emitters would be used for thinner materials and γ-emitters for thicker materials.

Radiotherapy

γ-Radiation from a strong source is used to kill cancer cells. Careful arrangements are made to concentrate the radiation on the cancer cells and not kill other healthy cells in the body. Extensive shielding is also needed to protect the medical staff operating the equipment.

Range and ionising effect of radioactive emissions in air

α-Particles move slowly, are charged and collide frequently with air molecules, which causes a very strong ionising effect. They lose energy in each collision, so have a short range.

β-Particles are also charged but move faster and collide infrequently with air molecules, so there is only a weak ionising effect. They lose energy slowly and have a range of a few metres in air.

γ-Rays move fastest, at the speed of light. They are not charged and rarely have collisions, so travel a long distance before losing all their energy. They have a very weak ionising effect.

Deflection of radioactive emissions in magnetic and electric fields

α-Particles are positively charged and are deflected as an electric current flowing in the same direction.

β-Particles are negatively charged electrons, so are deflected as an electric current flowing in the opposite direction.

γ-Rays are electromagnetic waves and are not deflected.

Examiner's tip

You need to be able to state and explain examples of practical applications of α-, β- and γ-emissions.

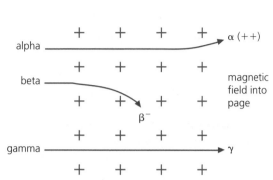

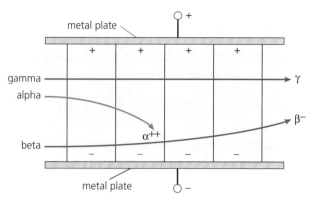

Figure 5.4a Deflection of α- and β-particles and γ-rays in a magnetic field

Figure 5.4b Deflection of α- and β-particles and γ-rays in an electric field

Now try this

The answers are given on p. 120.

4 A positron is a sub-nuclear particle with the same mass as an electron but with a positive charge. A certain nuclear reaction emits positrons and γ-rays, which are directed to pass parallel to and between two horizontal plates in a vacuum. The upper plate has a very high positive potential relative to the lower plate. Describe the path between the plates of:
a the positrons
b the γ-rays.

● Radioactive decay

Radioactive atoms have unstable nuclei and, when they emit α-particles or β-particles, they decay into atoms of different elements.

In α-decay, the nucleus loses two neutrons and two protons. The nucleon number goes down by four. The proton number goes down by two, so the nuclide changes to another element. Two orbital electrons drift away to make the atom electrically neutral and the new number of electrons equals the new proton number. An example of α-decay can be shown by a word equation:

radium-226 → radon-222 + α-particle

The same example of α-decay can be shown by an equation in nuclide notation:

$$^{226}_{88}\text{Ra} \rightarrow {}^{222}_{86}\text{Rn} + {}^{4}_{2}\text{He}$$

Note that, because an α-particle is the same as a helium nucleus, it is shown as 'He' in nuclide notation.

In β-decay, a neutron in the nucleus changes to a proton and an electron, which is emitted at high speed as a β-particle. The nucleon number is unchanged. The proton number goes up by one, so the nuclide also changes to another element. An electron is attracted from the surroundings to make the atom electrically neutral and the new number of orbital electrons equals the new proton number. An example of β-decay can be shown by word and nuclide equations:

carbon-14 → nitrogen-14 + β-particle

$$^{14}_{6}C \rightarrow ^{14}_{7}N + ^{0}_{-1}e$$

Note that, because a β-particle is an electron, it is shown as 'e' in nuclide notation, with a nucleon number of 0, because it has negligible mass, and a proton number of −1, because of its negative charge.

γ-rays are usually given off during both α-decay and β-decay, and can be added to the equations, for example:

$$^{226}_{88}Ra \rightarrow ^{222}_{86}Rn + ^{4}_{2}He + \gamma$$

● Please note

In the cases you will consider, α-particles and β-particles will not be given off in the same decay reaction.
α-Particles and γ-rays are possible together.
β-Particles and γ-rays are possible together.

● Sample question

Radioactive strontium-90 (Sr, proton number 38) decays to yttrium (Y), emitting a β-particle and γ-rays. Show this decay reaction as a
nuclide equation. [4 marks]

Student's answer

$$^{90}_{38}Sr \rightarrow ^{89}_{39}Y + ^{0}_{-1}e + \gamma$$ [3 marks]

Examiner's comments

The student has mostly got the answer right. The nuclide symbol for strontium is entirely correct, as are the symbols for the β-particle and γ-rays. The student has also correctly deduced that the proton number of yttrium is 39, one more than that of strontium. However, there is confusion in finding the nucleon number, which must stay the same in β-decay.

Correct answer

$$^{90}_{38}Sr \rightarrow ^{90}_{39}Y + ^{0}_{-1}e + \gamma$$ [4 marks]

Now try this

The answers are given on p. 120.

5 Radioactive uranium-238 (U, proton number 92) decays to thorium (Th), emitting an α-particle and γ-rays. Show this decay reaction as a nuclide equation.

● Half-life

Radioactive decay is a random process. It is impossible to predict when an individual nucleus will decay. However, *on average*, there is a definite decay rate for each isotope.

The decay rate is expressed as the half-life, which is the time for half the atoms in a sample to decay. As the process is random, there may be some small fluctuations, but, with a large number of atoms in even a small sample, the half-life is effectively constant.

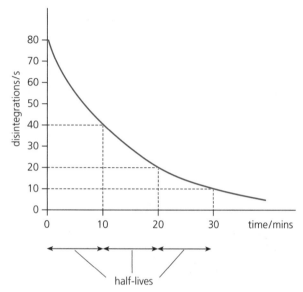

Figure 5.5 The half-life of a material can be found by using a graph (decay curve)

Examiner's tips

● None of the following have any effect on the half-life of a radioactive element: strong, electric, magnetic or gravitational fields, changes of temperature, aggressive chemical reagents or other elements the isotope is combined with.

● You must be able to understand and use the term 'half-life' in simple calculations with data in graphs or tables.

● Core students need to be able to understand and use the term *half-life* and carry out simple half-life calculations as indicated in Figure 5.5.

● Extended candidates need to be able to carry out half-life calculations allowing for background radiation, as in the following example. Be aware that background radiation also fluctuates.

● Sample question

A radioactive sample gives a detector reading of 700 counts per second, including a background count of 100 counts per second. The half-life of the sample is 7 days.

1 Work out the expected detector reading 3 weeks later.   [4 marks]

2 The value actually recorded 3 weeks later was 8 counts per second different from the expected value. Explain why this might be so.   [2 marks]

Cambridge IGCSE Physics Study and Revision Guide Second Edition © Mike Folland 2016

Student's answer

1 After 1 week, detector reading $= \dfrac{700}{2} = 350$ counts/s

 After 2 weeks, detector reading $= \dfrac{350}{2} = 175$ counts/s

 After 3 weeks, detector reading $= \dfrac{175}{2} = 87$ counts/s [2 marks]

2 The difference could have been caused by the sample warming up. [0 marks]

Examiner's comments

1 The student has failed to consider the background radiation, which does not decay.

2 The temperature of a sample has no effect on its half-life.

Correct answer

1 Initial detector reading due to sample = 700 – 100 = 600 counts/s

 After 1 week, detector reading due to sample $= \dfrac{600}{2} = 300$ counts/s

 After 2 weeks, detector reading due to sample $= \dfrac{600}{2} = 150$ counts/s

 After 3 weeks, detector reading due to sample $= \dfrac{600}{2} = 75$ counts/s [4 marks]

 Final detector reading including background = 75 + 100 = 175.

2 The difference could have been caused by a variation in the background radiation. [2 marks]

● Common error

✖ Stating that a radioactive sample loses the same number of atoms in the second half-life as in the first.

✔ If a sample of 10 000 atoms has a half-life of 1 hour, 5000 atoms decay in the first hour. In the second hour, 2500 atoms decay, that is, half of the 5000 that were left.

● Dangers of radioactivity

Radioactive emissions are ionising radiation. Ionising radiation can be very harmful to living things. It either kills cells outright or deforms cells, which can lead to cancers.

 Safety precautions:

● Whenever possible, radioactive samples should be in sealed casings so that no radioactive material can escape.
● Samples should be stored in lead-lined containers in locked storerooms.
● Samples should be handled only by trained personnel and must always be supervised when not in store.
● Radioactive samples should be shielded and kept at as great a distance as possible from people. In the laboratory, they should be handled with long tongs and students should keep at a distance. In industry, they are usually handled by remote-controlled machines.

- Workers in industry are often protected by lead and concrete walls, and wear film badges that record the amount of radiation received.
- Research, educational and industrial establishments using radioactive sources should have appropriate security measures in place to prevent abuse by mischievous, criminal or terrorist intruders. It is also necessary to have strict operational procedures to avoid accidental or deliberate misuse by members of staff.

● Sample question

An extremely strong source of α-particles and γ-rays is used in an experiment being demonstrated to a group of student observers. The source is held and moved by a robot arm controlled by a technician who is always at least 1 m away from the source. The observers are always at least 3 m away from the source.

1 These precautions are insufficient for the technician and for the students. Explain this. [2 marks]

2 Suggest practicable improvements that would permit the demonstration to continue and be observed in a safe way. [2 marks]

Student's answer

1 *The students would be safe at that distance but the technician needs to move further away.* [0 marks]
2 *The technician could use a video camera.* [1 mark]

Examiner's comments

1 This answer makes no distinction between α- and γ-radiation and fails to realise that 3 m of air will scarcely diminish the radiation from a γ-source.

2 The answer shows some awareness of the possibility of remote observation but does not address the issue of appropriate shielding.

Correct answer

1 The robotic handling and distance from the source protects the technician and students observers from the α-radiation. Because of the strong γ-radiation, even at 3 m distance, this is extremely unsafe for both the technician and the observers.

2 The source should be shielded by thick lead or concrete from all humans. The experiment could be viewed by video camera. A remote screen behind the shielding would allow the technician to control the robot and the students to observe in complete safety.

Answers

All numerical answers are quoted to 2 s.f.

Topic 1

1 a 9 swings.
   b 5.4 s.
   c 0.60 s.
2 a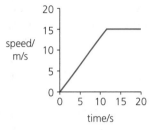

   b 210 m.
   c 11 m/s to 2 s.f. (10.5 m/s exactly).
3 a 0.80 g/cm³.
   b 2.8 g/cm³ to 2 s.f. (2.75 g/cm³ exactly).
4 a resultant force = 1800 − 1500 = 300 N.
   b mass $= \dfrac{\text{weight}}{10} = \dfrac{1000}{10} = 100$ kg

   acceleration $= \dfrac{F}{m} \dfrac{300}{100} = 3.0$ m/s²
   The rocket accelerates upwards at 3.0 m/s².
5 10 N down.
6 a momentum before collision
      = 6000 × 6 + 10 000 × 2
      = 36 000 + 20 000
      = 56 000 m kg/s
      momentum after collision = 16 000 × 3.5
                              = 56 000 m kg/s
      So momentum is conserved.
   b Reduced by 30 000 J.
   c Most of the energy is transferred to thermal
     energy in the buffers; a very small amount of
     energy is released as sound.

Examiner's tip

Think carefully about if the sound energy is significant.
This is a borderline case. There is a loud noise when
trucks collide but this actually requires little energy.
Marks will be scored when sound is a significant part
of the energy produced, e.g. from a loudspeaker.

7 a Gravitational potential energy.
   b Kinetic energy, elastic (strain) energy.
   c 5300 J to 2 s.f. (exact answer 5250 J)
8 a There is no air pollution, but the other types
     of pollution are a matter of opinion. Some
     people consider that the landscape can be
     spoiled (visual pollution), but others like to
     look at wind turbines. Wind turbines create
     noise pollution that upsets some people. The
     scientific evidence about whether or not noise
     is a real problem is unclear.

The running costs are low, as the energy
source is free, but wind farms are very
expensive to construct.
Over a whole year, a well-chosen location will
give a reliably predictable amount of energy.
But, day to day, the amount of wind is very
uncertain, so wind farms need to be part of
a complete energy strategy for a country,
together with other types of electricity
generation.
   b Wind energy will not be used up.
   c e.g. Energy from hydroelectric dams is from a
     renewable source.
     e.g. Energy from burning oil is from a non-
     renewable source.
9 a force acting down the slope due to weight,
     friction force (either way around)
     gravitational potential energy, work done
     against friction
     thermal energy
   b Friction in the bearings, electrical resistance
     of the wires.
     (A very advanced answer, which is not
     required at IGCSE but would earn a mark:
     heating of the soft iron core due to change of
     magnetic flux.)
10

Input energy/ working	Output energy/ working	Device
Chemical	Electrical	Battery
Electrical	Light	**Lamp/torch/ flashlight**
Kinetic	**Electrical working**	Wind turbine
Light (solar energy is too vague)	Electrical	Solar cell
Electrical working	**Sound**	Loudspeaker

11 76% to 2 s.f. (76.39% exactly).
12 a C.
   b B.

Topic 2

1 a When a molecule strikes the wall, its
     momentum changes direction from towards
     the wall to away from the wall. This requires
     a force from the wall on the molecule in the
     direction away from the wall. The reaction
     force from the molecule on the wall is
     towards the wall.
   b The pressure results from the total of the
     forces from all the molecules spread over the
     area of the walls.
2 Stage 1:
   a Increases: molecules always speed up with
     increases in temperature.

b Increases: molecules take less time to move from one wall of the container to another, so there are more collisions per minute.

c Increases: more collisions mean greater total force, so more pressure.

> Extended candidates should give extra explanation: greater rate of change of momentum requires greater force.

Stage 2:

a Stays the same: at the same temperature, the speed of molecules stays the same.

b Decreases: molecules take more time to move from one wall of the larger container to another, so there a fewer collisions per minute.

c Decreases: fewer collisions mean less total force, so less pressure.

> Extended candidates should give extra explanation: smaller rate of change of momentum requires less force.

3 The strips will curl with the bronze on the outside. The greater length of the bronze means it must be on the outside of a curve.

4 **a** 0 °C.
  b 100 °C.
  c 47 mm.
  d 70 °C.

5 **a** The sensitivity of the clinical thermometer is much higher, as the scale for each °C is much expanded and temperatures can be read to 0.1 °C. The thermometer has a very narrow capillary tube.

  b The range of the clinical thermometer is much less. It is reduced to 7 °C from about 50 °C, as only this range is relevant to the temperature of a human body. The clinical thermometer is shorter. In addition, its higher sensitivity means a reduced range.

  c Both thermometers have good linearity, as this is essential in any accurate thermometer. Mercury has regular expansion and the bore of the capillary tube is constant.

6 **a** 60 000 J.
  b 970 J/(kg °C) (exact answer 967.7 J/(kg °C)).
  c 130 J/°C (exact answer 134.61 J/°C).
  d 0.14 kg (exact answer 0.1391 kg, 0.134 kg if rounded answers from earlier parts are used).

7 **a** Stays the same: the ice is melting, so it stays at 0 °C.

  b Decreases: the latent heat to melt the ice is taken from the water, reducing its internal energy.

  c Increases: some of the ice has become water.

d Stays the same: the increase of mass of the water equals the loss of mass of the ice.

8 Differences: (1) Boiling takes place throughout the liquid. Evaporation takes place only on the surface. (2) Boiling takes place at one specific temperature but evaporation takes place at all temperatures. (3) Energy must be supplied continuously to maintain boiling. Energy is taken from the liquid during evaporation. Similarities: (1) Molecules move from being in a liquid to being in a gas. (2) Energy must be supplied to break the bonds holding liquid molecules together.

9 **a** Conduction: the metal of the gate is a good conductor, so thermal energy flows from his hands to the gate.

  b Convection: thermal energy flows from his hands to the water in contact with them, increasing its temperature. This water then moves away and is replaced by cold water.

  c Radiation: electromagnetic radiation transfers thermal energy from the fire to his hands, not requiring a medium between them.

Topic 3

1 **a** 0.75 m.
  b 0.20 Hz.

2 **a** Towards and away from him.
  b Vertically up and down.

3 **a** B.
  b E.
  c B = 30 °, C = 90 °, D = 71 °, E = 19 °, F = 90 °.

4 **a** An endoscope is used and inserted into the patient to reach the region to be examined. Light passes along the optical fibres (endoscope) to illuminate this region. Despite any twists and turns, total internal reflection within the optical fibre ensures that all the light reaches the end. Light from the illuminated region returns back in a similar way through other fibres and the doctor can see the region.

  b The signal is transmitted in digital form as a series of pulses of light. Light passes along the optical fibres to sensors at the receiving device. Despite any twists and turns, total internal reflection within the optical fibre ensures that all the light reaches the end. Any reply signal is returned in a similar way through other fibres.

5 a Real, inverted and magnified.
b Height 3.0 cm (reasonable tolerance 2.7–3.3 cm).
c 9.0 cm from the lens (reasonable tolerance 8.3–9.7 cm).
6 a The two rays diverge and will not meet at an image.
b (i) Virtual, upright and magnified.
(ii) Height 4.5 cm (reasonable tolerance 4.1–4.9 cm).
(iii) 6.0 cm from the lens (reasonable tolerance 5.4–6.6 cm).
7 a Upside down, so they appear the correct way up to the observer.
b enlarged, inverted, virtual.
8 a 5.5×10^{14} Hz (5.4545×10^{14} Hz).
Notes: Extended candidates are expected to know the speed of electromagnetic waves. All students are expected to be able to use standard notation, including positive and negative indices.
b 7.1×10^{-7} m
Notes: Students are not expected to know this value. However, they should be aware that the visible spectrum is a narrow range of wavelengths. Students should know that infrared has a longer wavelength than ultraviolet. It is a reasonable deduction that red has the longest wavelength of the colours.
9 a X-rays.
b Gamma rays.
c Same.
d 3.8×10^{13} km (exact answer 3.78432×10^{13} km).
10 28 m (exact answer 27.825 m).

Topic 4

1 a North.
b North.
2 a Copper.
b Soft iron.
c Hard steel.
3 a Charges like those on the plastic move to the far side of the stream, unlike charges move to the side close to the plastic.
b Like charges: weak repulsive force. Unlike charges: stronger attractive force.
c It is deflected towards the plastic.

4

Measurement to be taken	AB	CD	EF
e.m.f. of battery	Voltmeter	Nothing	Nothing
Current through *R* when connected to battery	Nothing	Nothing	Ammeter
p.d. across *R* when connected to battery	Nothing	Voltmeter	Wire

5 50 °C to 2 s.f. (50.29 °C).
6 a 4.4 V.
b 0.4 V.
c 1.0 V to 2 s.f. (1.015 V).
7 a 3 A.
b 2 A.
c 5 A.
d 1.2 Ω.
8 a 6 V (or just below 6 V).
b 0 V (or just above 0 V).
9

A	B	C	D	E
0	0	0	1	0
0	1	0	0	1
1	0	0	0	1
1	1	1	0	0

10 a The movement of the magnet causes the magnet field it produces through the coil to change. This change induces an e.m.f. across the coil. The e.m.f. causes a current to flow, which makes the coil an electromagnet.
b The direction of the induced e.m.f. and resulting current must oppose the change causing it. The bottom of the coil becomes an N pole. This pulls up the S pole at the top of the magnet to resist its downward motion.
11 4000 turns.
12 a The field direction alternates.
b The force direction alternates.
c The cone vibrates at the frequency of the a.c. signal.
13 a The coil will turn (or experience a moment or torque) clockwise.
b (i) Same.
(ii) Increased.
(iii) Decreased to zero.
(iv) Increased.
14 a (i) Up.
(ii) Down.
b (i) Decreased to zero, no current flows in the coil.
(ii) Decreased to zero, no current flows in the coil.
c (i) Down.
(ii) Up.

Topic 5

1

Isotope	Number of protons	Number of neutrons	Number of electrons
$^{88}_{38}\text{Sr}$	38	50	38
$^{90}_{38}\text{Sr}$	38	52	38

2 a Nuclear fission means splitting. The nucleus of an element with a high proton number is bombarded by a neutron moving with a suitable velocity. The nucleus splits into two nuclei with lower proton numbers.
Nuclear fusion means joining together. Two nuclei, usually of elements with a low proton number, join together under conditions of extremely high temperature and pressure. A nucleus with a higher proton number is formed.

b A large amount of energy is released.

c In the reactor of a nuclear power station.
d Inside the Sun.

3

Sample	CR (none)	CR (card)	CR (Al)	CR (Pb)	α	β	γ
1	6000	1000	1000	20	✔		✔
2	3000	3000	20	20		✔	

4 a Deflected downwards.
b Undeflected.

5 $^{238}_{92}U \rightarrow ^{234}_{90}Th + ^{4}_{2}He + \gamma$.

Cambridge IGCSE Physics Study and Revision Guide Second Edition © Mike Folland 2016

Index

Cambridge IGCSE Physics Study and Revision Guide Second Edition © Mike Folland 2016